4979

Farm Livestock

2010

Farm Livestock

GRAHAM BOATFIELD

Text illustrations by Keith Pilling

Farming Press

First published 1979
Third edition 1994

ISBN 0 85236 274 9

A catalogue record for this book is available
from the British Library

Published by Farming Press Books
Wharfedale Road, Ipswich IP1 4LG, United Kingdom

Distributed in North America
by Diamond Farm Enterprises,
Box 537, Alexandria Bay, NY 13607, USA

Cover design by Liz Whatling
Typeset by Galleon Typesetting
Printed and bound in Great Britain by Page Brothers, Norwich

Contents

There is a colour section between pages 84 and 85

Illustrations

Acknowledgements

My grateful thanks are due to my professional colleagues from various Colleges of Agriculture who have helped with the supply, the processing and the preparation of material used in this book. These include Mr Ian Hamilton, formerly of the Hampshire College, and Mr Chris Keeble of the Otley College.

My friend Mr Philip Wood and my son Mr Laurence Boatfield have both provided useful guidance and scrutiny in the revision of the text. I should also like to thank the many readers who have provided comments on earlier editions of the book.

Introduction

THIS BOOK is designed to give a simple outline of farm livestock, their feeding, breeding, health and management, as found on British farms today.

The course of study set out here needs positive work by the student, based on farm observations and simple studies carried out locally. A book of this type gives a very general picture—local information is vital to complete that picture. Animal husbandry and livestock farming vary according to local needs and systems of farming. National averages and general comments mean little if they are not related to local conditions.

This book is in seven sections. The *Cattle* and *Sheep* sections cover the normal breeding, production and husbandry processes involved in the farming of these ruminant animals. The *Pig* section deals with these more intensive animals in the same way.

The chapter on *Feeding Livestock* describes general principles and practice, including the make-up of farm foods, the function of the main types of digestive systems and the rationing of farm animals.

Breeding Livestock deals with the reproductive system, the science of breeding and animal growth.

The section on *Animal Health* does not aim to rival a veterinary textbook, but to indicate the main concerns of the stockman and farmer in the care of animals on the farm. *Stockmanship* draws together many of the interests of the conscientious person looking after livestock.

Some current information is given by production targets, average results and standards, but it must be remembered that these change as farm practice improves. It is important to keep up to date with modern thought, local farming practice, and the knowledge that comes from the advisory services and the technical agencies. There is always plenty of good advice available to farmers and people in charge of livestock from the Ministry of Agriculture (ADAS), from the Meat and Livestock Commission (MLC), from the Milk Marketing Board (while it exists in its present form), from commercial firms and from veterinary surgeons. You will have to pay for some of this advice.

Farm Livestock is specially prepared to meet the needs of students, those new to farming and anyone who wants to understand modern farming practice.

The material presented covers the requirements of students taking courses which lead to national vocational qualifications. It is of value for school work in land-based subjects and rural and environmental studies, and for the private study of anyone with an interest in farming.

A useful annual reference is John Nix's *Farm Management Pocketbook*, published by Wye College (University of London), Wye, Ashford, Kent TN25 5AH.

Any references to farm safety and to animal welfare or other regulations and codes of conduct are given here purely as a guide. The reader should refer to the relevant leaflets and other publications to find the exact requirements of current regulations.

Figures are stated in both metric and British (imperial) terms for the convenience of readers. Figures and dimensions are adjusted conversions and not necessarily exact.

RARE BREEDS OF FARM LIVESTOCK

The work of the Rare Breeds Survival Trust is of particular interest. Since 1900 in Great Britain some 20 breeds of farm livestock have disappeared. Since 1973—when the R.B.S.T. started work—no British breeds have disappeared and now about 50 rare and less usual—sometimes called minority—breeds are supported and encouraged by the Trust. Some breeds, such as Jacob and Black Welsh Mountain sheep, have even become more numerous and well established again.

Fashions in livestock change. Sometimes breeds become scarce, or even disappear, because there seems to be no further need for them. But sooner or later there may be a demand for an old breed. Sometimes it is because certain qualities are required—such as resistance to temperature or special conditions—or because the type is needed for a breeding programme. At the moment, lean meat is preferred to fat; light skin in pigs is preferred to dark skin; horns are not favoured in cattle kept in yards or buildings; sheep wool is not so valuable as it once was, and sometimes hardly covers the cost of shearing.

All or any of these farming needs may change. Intensive farming of another sort of livestock may come in; environmental or 'green' considerations may become more or less important than they are at present. For various reasons, not always obvious, some people like to keep animals of an old breed, which may be rare. There is no reason why they should not.

Anyone who wants to support the work of the Rare Breeds Survival Trust should get in touch at the National Agricultural Centre, Kenilworth, Warwickshire CV8 2LG (tel: 0203 696551) or with one of the local support groups of the Trust found in many parts of Great Britain. There are similar organisations in some other European countries, and in the United States.

OTHER ANIMALS

This book deals with the main types of livestock commonly kept on farms in western Europe. Tradition often decides what animal products are in demand in one area or another; examples of this include goats' milk, duck and goose liver, and horse meat. With increasing trade between countries, the unification of Europe, and a more cosmopolitan approach to shopping and foreign travel, public taste changes and the demand for new animal products widens. Here are short notes on some 'new' animal types on farms, and products which are found increasingly in shops and markets.

Deer

These ruminant animals produce the high value meat venison, which is increasingly found regularly in shops, rather than as an occasional product according to the vagaries of the season (rather like wild fungi). Deer are now being farmed in larger units, needing high fencing and secure conditions. Reindeer also are farmed in parts of Scandinavia and on a small scale in Scotland.

Fish

Fish farming, which was common in the Middle Ages, is now increasing in size and scope in many countries. It is a very intensive and specialised form of farming, with its own problems of marketing, international trade, local demand and pest and disease control.

Goats

Goats have traditionally been kept in small numbers attached to the household or in larger numbers in poor countries and in hot climates. For reasons of health and changing public taste, goats' milk is in demand for some people with special needs, and for making specialist cheeses, often of high quality. Larger goat units are now being set up, both for milk production and for the supply of goat flesh from young animals. There is also a market for fibre.

Horses

Fewer working horses are found on farms today, but as standards of living rise there is a greater demand for riding horses, often in small units and with intensive use of grazing. In Britain this sometimes goes by the slightly derogatory name of 'horseyculture'. Horses (and asses in some countries) can be used for meat production, but this is a practice which arouses some strong feelings and is restricted to certain areas.

Poultry

Hens and their eggs are found everywhere, but local preferences and prohibitions apply. The general tendency is toward large units and the 'factory style' of production has so far prevailed, although a demand for so-called 'free range' (more extensively kept) poultry is also found.

In some countries there is an expensive market for the livers of specially fed ducks and geese. The Barbary duck, usually known as the Muscovy in Britain, is now found in shops and markets. The ostrich has recently been introduced from Africa as a meat-producing animal.

Rabbits

Rabbits have traditionally been kept in small units, often behind the family house, but are now raised in purpose-designed buildings in larger, more intensive units. Rabbit meat, fresh and frozen, is now more often in shops. The digestive system of the rabbit is different from other farm animals, and is briefly described on page 75.

Snails

Snails were originally gathered, seasonally, from the wild, and in special areas such as vineyards, there is an increasing tendency to 'farm' snails, and to sell them prepared for the kitchen in supermarkets and other shops. There are two main types of edible snails: the large (Roman or Burgundy snails) and the smaller (little grey) snails.

Wild Boar

Wild boars were formerly kept in woods and forests, mainly for hunting. Units are being set up both for breeding pure-bred animals and for crossing with ordinary farm pigs, and the meat is now more readily available. Regulations concerning wild and/or dangerous animals may apply to them.

Chapter 1

Cattle

CATTLE ARE the most productive farm livestock kept in the United Kingdom; they produce not only large quantities of milk but also meat of different types at various stages in their lives. During recent years there have been great changes in cattle as a result of breeding and management. There is a wide difference between types, some breeds have become very much more important than the rest, and new methods of beef production have become common.

CATTLE FOR MODERN CONDITIONS

Milk must be produced to a high standard of health and hygiene, and contain definite amounts of the various food substances. It must be produced as cheaply and efficiently as possible. In addition to good ordinary milk which is up to standard, there is also a demand for 'Channel Island' milk, which is richer and sells at a higher price; for skimmed milk, from which the fat is partly removed; and for other specialised types of milk.

Beef, like other meat, must not be too fat; it must be lean and tender, and suitable for modern packaging and selling. Beef is now being produced from younger animals, as a very intensive form of livestock farming.

Types of Cattle

Cattle are kept under very different farming conditions, and no one type or breed is suitable everywhere. Although there are distinct beef and dairy types, there is no clear line between them. Three-quarters of our beef is produced from the dairy herd. Our chief dairy breed, the Friesian, is a good beef producer, but the Holstein type is less 'beefy' and takes longer to finish.

- *Dairy type* The extreme dairy-type of cow is long and thin, with a wedge-shaped body, carrying little flesh and with prominent bones.
- *Beef type* A real beef-type animal is solid in its body, well covered with flesh, particularly in the hindquarters and back.

1

In the past, there were several breeds of a so-called 'dual-purpose' type cattle, producing both milk and beef but sometimes doing neither job very well. Such breeds are of little importance today in Great Britain; some have, in fact, become quite rare.

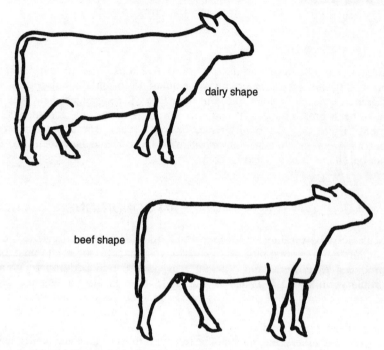

Figure 1   The dairy shape and the beef shape

SIGNS OF HEALTH

These are some signs which show whether cattle are in good health and condition:

- Bright eyes, loose skin, lick marks on coat, good appetite, chews the cud, dung and urine normal, moist nose, ears and base of horns warm to the touch. Holds its head well, walks properly
- Watch changes in milk yield and general behaviour—which may be due to illness
- Temperature: 39°C (102°F)
- Pulse rate (taken under the base of the tail): 60 per minute

How to Tell the Age of Cattle

There are 8 front teeth in the calf's lower jaw. These are replaced by larger permanent teeth (broad teeth), from the centre first, as follows:

2 broad teeth $1\frac{1}{2}$ years old
4 broad teeth $2\frac{1}{4}$ years old
6 broad teeth $2\frac{3}{4}$ years old
8 broad teeth $3\frac{1}{2}$ years old

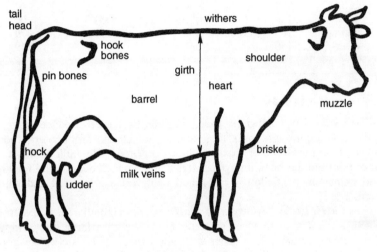

Figure 2   The points of a cow

CATTLE BREEDS TODAY

Since the beginning of this century there have been great changes in the relative importance of different breeds and types of British cattle. Then, the most important was the Dairy Shorthorn. Like some other breeds which have become less important, it was one of the 'dual-purpose' breeds, aiming to produce both milk and beef.

Over the years the British Friesian, a breed brought originally from Holland, has become dominant; today there are more cows of this breed than of all the others put together. It is kept mainly to produce plenty of milk, but it also produces useful bull calves for the type of beef production that suits modern needs. A large proportion of our home-produced beef is from Friesian cattle (see British Holstein below).

Only about one-third of our cattle is bred purely for beef production. Bulls of these breeds are used widely for crossing with other breeds for commercial beef production.

The main dairy breeds in England and Wales are as follows:

Breed	Liveweight (kg)	Per cent of dairy (type of cow)	Milk yield (kg) per year	Butterfat per cent
Friesian	560	74.50	6087	3.97
Holstein	560	3.40	6825	3.94
Ayrshire	460	1.90	5453	4.00
Shorthorn	500	0.60	5327	3.79
Guernsey	450	1.40	4569	4.79
Jersey	380	1.70	4262	5.45

Notes on Our Breeds

Friesian Our chief dairy breed, popular because it produces a lot of milk. It is a quick-growing, large animal, which can produce good lean meat, either pure-bred or crossed with a beef breed. The milk quality was once poor but has been improved. It is a breed which needs good feeding and management. Colours: black and white (also a few red and white herds).

British Holstein A strain of the Friesian, originally imported from North America, selected to produce animals with a large bony frame. The carcase remains lean even with good feeding. Holsteins are good users of bulky and fibrous foods, for high milk production. Colours: black and white.

Ayrshire A hardy and healthy breed from Scotland, produces good quality milk and will do well under poorer conditions. A very neat type of cow, of good dairy shape. Not suited to early beef production. Colours: mixtures of white with red or brown.

Shorthorn Once our chief breed for milk production, it has now declined in numbers. It was kept to produce both milk and beef. Colours: red, white, red and white, roan.

Guernsey A Channel Island breed, giving rich milk. Scattered in England, but kept mainly in the South. Very much a dairy type of cow. Colours: white with brown or fawn, in patches.

Jersey A Channel Island breed, giving very rich milk. Widespread in England, kept mainly in the South. An extreme dairy type of cow. Colours: fawn, to dark brown or grey with black muzzle and tail.

Red Poll A dual-purpose breed, which started in East Anglia and became widespread, but is now seldom found.

South Devon The heaviest British breed. It produces milk which is up to Channel Island quality, and good beef. Colour: light brownish-red.

Welsh Black A hardy breed suited to poor conditions, used for beef production. Colour: black.

Hereford Our chief beef breed, popular for crossing with many other breeds. Hardy grazing cattle. Colours: red, with white face, chest and underline.

Aberdeen Angus A very high quality beef breed, which needs good feeding to do its best. It is short-legged, early maturing, with a high proportion of meat to bone. Colours: black. No horns.

Galloway A hardy grazing breed, suited to poor conditions. Colours: black or dun. No horns. (There is also a Belted Galloway, with a white belt around the body.)

Lincoln Red It is really a red type of Shorthorn, found mostly in the East Midlands. Now used for beef production. Colour: red.

North Devon (Often called Devon.) A good quality, quick-growing beef breed. Colour: ruby red.

Sussex A beef breed, very similar to the North Devon.

Highland a very slow-growing, shaggy, beef animal, suited to very poor and hard conditions. Colours: various shades of brown and dun.

Imported Breeds

Several breeds of cattle, from other European countries and elsewhere, have been brought into Great Britain for various purposes and some of them have now become established. They are sometimes known as exotic breeds.

Charolais First imported from central France in the sixties for experimental crossing with some of the dairy breeds, it is now established in Great Britain and other countries. It is a large, heavy breed, especially well fleshed in the hindquarters, with a limited amount of fat. Colours: white or cream only.

Limousin Imported from central France, it is a very large, hardy breed, similar in some ways to the South Devon. It has good carcase quality, growth, and high killing-out percentage (see page 23). They grow quickly and live long. Colour: light brownish-red.

British Simmental Came from Switzerland originally and is widely kept in Germany and France. It is a breed which produces both a good milk yield and when reared for beef, plenty of lean meat. In Britain, it is used for crossing. Colours: white face, yellowish brown or red with white markings.

Blond D'Aquitaine and *Salers* both come from France. Other imported breeds may appear from time to time.

CATTLE BREEDING

A young bull may be used for breeding from the age of 12 months. Using natural service, a young bull can be mated to 20 cows per year, the number increasing as it gets older. Bulls are dangerous animals, although the bulls of some breeds may appear placid and safe enough, and no one should take any chances with them. In practice now, most British dairy cows are mated by artificial insemination (AI). (See page 100.)

The heifer or cow will come 'on heat' every 20 days until she is 'in calf'. After calving, it is some 5 to 8 weeks before the cow comes 'on heat' again.

A heifer (so called until it has had its second calf) is first mated at an age which varies according to breed and to systems of management; 15-18 months is a normal age now for the first mating.

The heifer or cow is 'in calf' for $9\frac{1}{2}$ months (average 285 days), and the aim is to produce a calf each year. The time of year when calves are wanted varies with the system of farming: autumn calving is usual in milk producing herds, and spring calving common in beef breeding herds.

Signs of Coming on Heat

- The heifer or cow becomes more excitable.
- She allows herself to be mounted by other cows.
- Her temperature rises for a time.
- Milk yield drops slightly.
- There is sometimes a clear discharge from the vulva.

Mating

The cow (or heifer) is on heat for about 24 hours, and it is best if mating is done towards the end of this period. With natural mating, a bull may be run out in the field (or in a yard) with the herd and serve cows which are on heat, when they are ready to stand. With a dairy herd, it is more common to house a bull in a special pen, and bring cows to him when they are ready to be served. However, most mating is done now by AI. Bulls of various breeds are kept at a centre, and their semen is taken and brought onto the farm by an inseminator who puts it into cows which are ready for mating. It is more common today for the insemination to be done by the person in charge of the cattle.

The heifer or cow must be fed well during the time it is 'in calf', and with dairy cattle it is common to 'steam up' (feed well to encourage future milk production) during the last 6 weeks before calving.

Signs of Calving

- The heifer or cow acts in a nervous or unusual way and becomes restless.
- The udder stiffens, gets much larger, and fills with milk.
- The pin bones widen and the muscles slacken on either side of the tail.
- The vulva swells and becomes much larger, and there is a discharge.

The cow will usually calve without any help, but sometimes she needs assistance and in bad cases it is a job for the vet. The new-born calf may need a little attention: remove the membranes from around the head, clear its mouth, pull the tongue forward, rub down with straw, and if it is not breathing, work the chest with the hands. Let the cow lick the calf if they are to stay together, but if the calf is to be taken away do not let the cow see it at all.

You may have to dress the navel of the calf with an antiseptic to avoid navel ill—a type of infection.

CALF REARING

The colostrum (sometimes called 'beestings' in cattle) is important to the calf and it should have a supply for the first 3 or 4 days of its life. This 'first milk' of the cow gives the calf some protection against common diseases of the early days; it also contains extra protein and vitamins, and has a necessary laxative effect.

Calves are usually born singly. Twins are uncommon (one in every 80 calvings) and triplets or more are very rare. A bull calf is normally heavier than a heifer calf, and very often is born later than expected.

These figures show the average weights of calves at birth:

Breed	Heifer calf	Bull calf
Jersey	25 kg (56 lb)	27 kg (61 lb)
Ayrshire	30 kg (66 lb)	34 kg (76 lb)
Friesian	39 kg (86 lb)	42 kg (92 lb)

A calf is not as hardy as a cow and for the first few months needs special care and management.

The calf's stomach is at first simple. It cannot 'chew the cud' like the cow, and for this reason cannot deal with much fibrous food (hay or straw) nor with much bulk food. As the calf grows its stomach develops and later in life it can deal with all sorts of food, and plenty of it.

Whatever the system of rearing, the calf needs colostrum at first, either direct from its mother or fed from a bucket. After this it will have milk, fed in various ways, or substitute foods.

Natural Rearing

The calf gets milk direct from a cow. There are two main methods:

Single suckling   One calf to one cow, usually suckling for 6 months, after which it is weaned. This is a rearing method used on poor land with cows which do not give a lot of milk.

Multiple suckling   Several calves to a cow. It is common to use 'nurse cows', which are often old dairy cows with some faults. One method is to put 4 calves on the cow for about 10 weeks, wean them, then 3 more calves, for another 10 weeks, followed by 2 or 3 calves, and after that 1 or 2, according to how much milk the cow is giving.

Artifical Rearing

The calf is kept away from the cow and is fed separately. There are many different methods.

There are liquid feeding systems using cold colostrum (stored in the freezer) or acidified milk and an increasing use of automatic feeders.

Whatever system is used, the calf should not have too much milk *in its first few days*, as this can upset it. A good rule is to feed 10 per cent of its birth weight: 4 kg milk per day for a 40 kg Friesian calf and 2.6 kg (or just a little more) for a 26 kg Jersey calf.

Most calves are reared today by *early weaning* methods, given *colostrum* for 3–4 days only, followed by milk substitute for 5 weeks (or even as little as 3 weeks). Whole milk is too expensive to feed to calves today, and is only likely to be given in any quantity for such special purposes as the rearing of high quality bulls. The calf is encouraged to eat dry concentrate foods as early as possible. Hay or straw is put before the calves, so that they can help themselves. Always give calves good quality hay or straw.

Daily Rations at Various Ages

Age	Milk (or milk substitute)	Concentrates
1 to 4 days	Colostrum	—
Up to 10 days	Up to 4 litres	Ad lib
Up to 4 weeks	Up to $4\frac{1}{2}$ litres	Ad lib
5th week	Reduce and then stop	Ad lib
Up to 12th week	None	Up to $2\frac{1}{2}$ kg

Note: Whatever milk substitute is used, mix according to the maker's instructions.

Health

You must take no chances in the artificial rearing of calves. It is usual in Britain to expect a 6 per cent death rate, mostly in the first 3 weeks of life. Common troubles are:

Joint-ill An infection of very young calves, through the unhealed naval. Prevented by hygiene and by dressing the navel and cord with antiseptic at birth.

Scours Forms of diarrhoea, due to faulty feeding, poor management or infection.

Pneumonia Serious infection of the lungs, caused by wet, cold, draughts and infection.

Rules of Calf Rearing

- The calf should have colostrum whenever possible for the first few days.
- Never feed too much.
- Let the calf have plenty of clean water.
- Make any changes in the ration gradually, and only feed good quality foods.
- If there are any signs of scouring, stop all food and treat with great care.
- Keep all buckets, troughs, etc., very clean.
- Keep the calf pens clean, light and airy.
- Treat calves quietly and carefully at all times.

Rate of Growth

We neither need nor expect all calves to grow at the same rate. Their rate of growth varies according to their purpose later. It certainly pays to treat calves and young stock well (this applies to all young animals) but dairy heifers, for example, do not need to be raised in too good a condition. Young beef cattle (like young pigs) need to be pushed on for quick growth. These figures give some idea of rates of growth to aim for:

	Liveweight increase per day in kg
For veal production	1.0
For bull rearing or beef	0.9
Larger breeds to be reared for dairying (e.g. Friesian)	0.7
Smaller breeds to be reared for dairying (e.g. Jersey)	0.5

Some Routine Jobs

Various things need to be done to calves during the first few months.

Dehorning   Cattle are better without their horns, to avoid damage to people and other animals. This can be done by using a hot iron (first injecting an anaesthetic).

Castration   It is possible to rear male calves for young beef production (up to 12 or even 18 months old) without castrating them, but most male calves are still castrated, if they are not to be kept for bull rearing, by:

- *The knife*   Surgical removal of the testicles.
- *Bloodless*   Using the Burdizzo castrator (read the instructions first).

Rubber rings are *not* recommended for calves.

Removing extra teats   Extra teats should be snipped off with a sharp pair of scissors during the first month.

Calf Housing

Most calves, unless running with their mothers or with nurse cows, should be kept in special pens. It is wise to keep them in individual pens for the first month, possibly longer. Housing calves separately prevents them from sucking one another, and cuts the risk of disease. Later, calves of about the same size can be run together in groups.

Buildings for housing calves should be:

- Draught free
- Light and airy
- Easy to clean
- Well laid out for convenient working, storage and handling of materials

There is no point in having a calf house hot (like a piggery), and a temperature of 10–13°C (50–55°F) is warm enough. The important thing is to avoid rapid changes of temperature. Very young calves can be warmed by putting them under an infra-red lamp.

It is very important to avoid the spread of disease, and a thorough disinfection of the whole building and all fittings, at least once a year, is a great help; if possible, rest the building from calves altogether for a while.

Many people like to give calves a good bed of dry straw to lie on, although some find it best to give very little bedding.

Special fittings are used to hold feeding buckets outside the pens (sometimes with a 'yoke' arrangement to hold the calves by the head); hay or straw is given in a rack or a net of about 5 cm (2 in) mesh).

	Age of calf	Area per calf
Individual pens	To 4 weeks	1.5 m by 0.75 m
	To 8 weeks	1.8 m by 1 m
In groups	To 8 weeks	1.1 m²
	To 12 weeks	1.5 m²

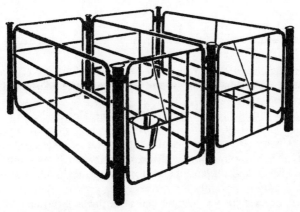

Figure 3   Typical calf pens

Management after 3 Months Old

After this time the calf will be able to deal with more bulky foods. It should have the best quality hay at a daily rate of $\frac{1}{2}$ kg for each month of age (e.g. a 4-month calf should have 2 kg per day) and good silage, or kale or roots, with concentrates.

Autumn-born calves should grow well through their first winter and go out to grass when it is growing well in the spring. Many spring-born calves do not go out to grass in their first summer, but are kept in and fed well. A modern practice is to turn them out to grass earlier.

THE DAIRY FARM

The number of dairy farms is getting less all the time and those that remain have herds of cattle which are growing larger. Farmers with small herds of milking cows are leaving milk production; it does not pay on a small scale. These figures show how the average size of herd has changed in recent years, but there is of course a great difference between various parts of the country.

Year	Average number of cows per dairy herd	
	England and Wales	Your region
1960	21	
1977	58	
1984	67	
1986	68	
1991	71	

You can get figures for your region from the Statistic & Forecasting Department, Milk Marketing Board, Thames Ditton, Surrey KT7 OEL (telephone: 081 398 4101).

Most milk producers sell their milk (wholesale) to the Milk Marketing Board. Less than one in 22 (producer-retailers) sell direct to the housewife. Very few turn their milk into butter or cheese on the farms, and this only in some parts of the country (such as Somerset and Cheshire). Most milking is done with milking machines, and very little by hand.

Most cows are kept for commercial milk production; less than one in five for beef rearing. Most dairy farms do not keep a bull, and two-thirds of mating is done by artificial insemination (AI).

The Dairy Cow

The milking cow does a very hard job, and to do it properly needs correct preparation, care, and management. As a heifer she will be growing, as well as producing a calf and getting ready for her future milk production. Both heifers and cows are 'steamed up'—given extra food from 6 weeks before calving until the time they calve.

Just before calving, the udder builds up and fills with milk. This first milk is colostrum (see page 7) which is not fit for sale for the first 4 days after calving. The milk yield increases during the first few weeks after calving, until it reaches *peak yield*, which is held for 1–2 months. The whole period of milk production is known as a *lactation*; it usually lasts over a 10-month period, and the cow is dried-off (milking stops and she is given a rest) for 2 months, to prepare for the next lactation.

The cow does not produce milk without having a calf, and the aim is to calve her once a year, at a time to suit the system of farming. During the last few years winter milk has paid best and most dairy cows have calved in the autumn and winter, but in some grassland districts summer milk may be more profitable. To get the cow in calf again, and ready for the next lactation, she must be mated 2–3 months after each calving.

The graph below gives a picture of the average lactation of a cow. Cows vary, of course, which may be due to their own make-up and breeding, and also to management and feeding.

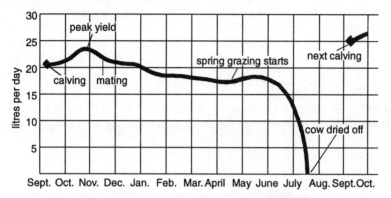

Figure 4 Graph of a cow's lactation

The cow's lactation produces about 25 per cent more milk than the same animal as a heifer. There is an increase in each lactation up to the fourth or fifth, which is the highest in a cow's life, and after that it gets less each year. We expect a good dairy cow to come up to these standards:

- She must give a good yield of milk. How much depends on the breed, the type of farming and the management.
- Her milk must be of good quality (see page 19).
- She must calve regularly.
- She must live long—so that the cost of her replacement can be spread over as many lactations as possible.
- She must be healthy and active.
- She must make the best use of her food.
- She must produce a good calf and breed true.

Systems of Dairy Farming

How cows are kept, and how they are milked, depends upon the type of cow, the type of land, the buildings, and the management—of which feeding is a most important part. A dairy farmer has to make a number of decisions. These are some of the choices:

- *Self-contained herds*   Replacements for the dairy herd are bred on the farm; calves are kept and reared.
- *Flying herds*   No replacements are reared on the farm, but in-calf or freshly calved cows or heifers are bought into the milking herd.

Winter or summer milk   There is a slightly higher yield from cows which calve in the winter months, and as the price of winter milk has been higher during recent years, it has paid on many farms to follow this policy. It is more common in the east of England than in the grassland districts of the West, where summer milk is sometimes produced from cows calving in the spring.

Feeding   There are two extreme methods:

- *Self-sufficiency*   This means producing as much food as possible on the farm. As the cheapest foods for cattle are grass, silage, hay, kale and roots, it means feeding the cows on as much of this bulk food as possible. Usually it means lower milk yields, produced at a lower price.
- *Heavy compound feeding*   The object is to produce as much milk as possible by feeding the cow well on compounds (see page 69) for milk production. Bulk foods will be used in moderation; some hay at least is always needed for cows. This system means high milk yields, produced at a high price.

At the present time, the aim is to produce milk as cheaply as possible and to avoid over-production. Expensive methods do not pay—and production of milk from home-grown foods is usually the cheapest method.

Housing   On light land, in sheltered districts, it may be possible to keep cows out all the time, but this is not common. Most cows are housed during the winter, and are now kept in yards of one sort or another—and milked in milking parlours.

Cowsheds were in use for many years and are still found on a number of smaller farms. The cows are tied up in standings, with a manger and feeding passage in front of them, and a dung channel behind them. It is not a labour-saving system, but it does allow individual attention.

Yard and parlour is the current system found on many farms and particularly with larger herds of cows. It means keeping the cows (which should have no horns) in the yards, and bringing them through a parlour for milking. There are many different layouts of yards possible, with *loose housing* of cows in some so that they run together in groups. Other systems use *cubicles* in which cows can lie individually, and come and go as they please into a yard. Another method is to keep cows in kennels, which are roofed.

The system used on any farm depends on a number of things: what buildings are already available; whether straw for bedding is short or plentiful; and how much money can be spent on buildings. Manure may be handled in the form of slurry (a semi-liquid form of dung) or as solid manure with straw.

Feeding systems also vary. A silage-feeding system may be in use; chopped maize, forage harvested crops or grass may be fed. A trough system may be used for feeding concentrate foods, and there may be yokes to hold the cows by the head while they are fed. Where feeding takes place may also vary: it may be done entirely in the yards. Some concentrates may be fed in the parlour, or the whole ration of concentrates may be fed, individually to each animal. There are elaborate types of equipment installed for this purpose.

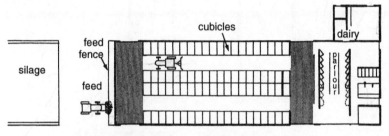

Figure 5   Yard and parlour layout

Parlour systems vary greatly; methods and fashions change from time to time, and what is suitable for one size of herd may not be right for another. Generally, systems are designed to deal with the cows as quickly as possible, and with the least number of workers. For smaller herds, there are *abreast* parlours with the cows side by side and *tandem* parlours with the cows standing in line. It is usual now for the operator to stand in a pit at a lower level than the cows, for ease of working.

Larger herds now are often milked in *herringbone* parlours, which allow for efficient working, with equal numbers of cows on each side of the pit. The most complicated and expensive system is the *rotary* parlour, with the cows standing on a rotating platform and the operators in the middle.

Milking equipment has also changed over the years. Cows were milked with bucket plants, carried to a central point; then pipeline systems were developed, so the milk went straight to a cooler. Now the cows are milked in the parlour, and the milk passes via the pipeline to a bulk milk tank where it is cooled by refrigeration. During the last few years, the bulk milk tank has replaced the churns which used to be

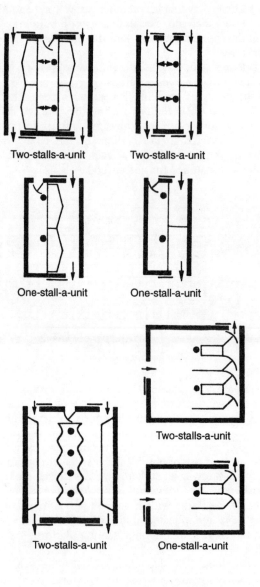

Two-stalls-a-unit

Two-stalls-a-unit

One-stall-a-unit

One-stall-a-unit

Two-stalls-a-unit

Two-stalls-a-unit

One-stall-a-unit

Figure 6   Parlour arrangements

collected by lorry; now the tank is emptied direct into a road tanker.

In the parlour, cows are cleaned and washed, and often washing systems are installed which supply warm water conveniently for the operator, with paper towels to dry the udders. Feeding systems in the parlour can be very complex, so that rations are supplied according to the needs—and the milk yield—of individual cows.

GETTING MILK FROM THE COW

Milk is formed in the udder from food substances carried in the blood stream. It is being formed in the udder all the time between milkings. To get as much milk as possible from the cow, it is important to milk her out quickly and completely.

A small amount of milk is held in the 'cistern' above each teat, but most is held in the tissue of the udder, and released as milking is done. The process of milking brings milk from the cistern into the passage through the teat, and from there out of the teat.

Let-down of milk is most important. The cow can release the milk from the udder tissue, or she can hold it up. Any form of nervous strain will make the cow hold up her milk—noise, strange surroundings, a change of cowman, or a change of routine.

Let-down can be encouraged by good milking technique, and in

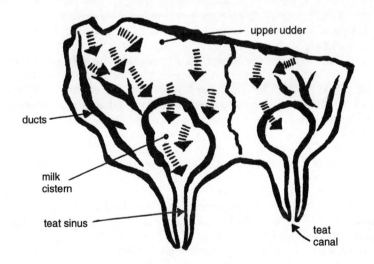

Figure 7   Inside a cow's udder

particular by preparing the cow properly beforehand. The udder should be stimulated by washing with warm water, milking should start within two minutes after this, and the whole job of milking should be finished as quickly as possible.

Stripping means getting the last few drops of milk from the teats. It can be done by machine—pulling down on the teat cups and massaging each quarter at the same time, but should not be prolonged.

There are 4 'quarters' to each udder—a quarter is one teat, and that part of the udder which lies above it, and which supplies it with milk. The two back quarters produce slightly more milk than the front two.

Most dairy cows are milked twice a day; three times a day milking is done with heavy yielding cows, and particularly in pedigree herds where it is important to get the highest possible yields. The first milk drawn from the udder (the 'fore-milk') should be drawn into a 'strip cup' and examined for signs of mastitis or other trouble.

However cows are milked, it is ideal to have the same interval between milkings—12 hours apart with twice-a-day milking. This is not usually possible, but it is important to keep to regular times.

CLEAN MILK PRODUCTION

There are various laws and regulations which concern the production of milk, and which exist to protect the public. Milk is a food which must be kept clean and healthy, as it can easily pick up and carry infection, especially to young children.

Registration   The farm must be approved for milk production by the Ministry of Agriculture.

Milk and dairies regulations   The farm and the methods of milk production must satisfy these regulations (see the booklet 'A Guide to Clean Milk Production' from the Ministry of Agriculture). Walls and floors must be impervious and easily cleaned, drains must discharge outside the buildings, there must be plenty of light, ventilation, and a pure water supply. There must be a separate milk room.

Health   Cows must be free from any sign of disease. They must be certified free from the disease tuberculosis (shown by tuberculin testing).

Central testing   Since 1984 all milk is tested weekly at regional laboratories. It is tested for compositional quality (butterfat, protein and lactose); for hygiene by TBC (total bacterial count); and for antibiotics. Payments are based on these tests.

Antibiotics   If a cow is treated with antibiotics against disease, a period must be allowed before the milk from the cow is sold.

Cool milk   The cooler the milk, the longer it keeps fresh. By law it must be cooled, and it is better to get it down to 4°C (40°F) as soon after milking as possible.

Clean cows Before milking, make sure the cows' udders are thoroughly cleaned, and washed with warm water and then dried.

Cleaning Milking Equipment

Cold wash Rinse everything thoroughly in cold water, to get rid of any traces of milk.

Hot wash The equipment is washed thoroughly in hot water with detergent in it.

Sterilise This can be done by the use of steam or hot water with hypochlorite in it.

Rinse After the hot wash with hypochlorite, rinse in water to remove traces of chemicals and detergent.

Special methods With the more complicated layouts in use today, the *circulation cleaning* method uses a pump to clean all parts of the system.

MILK QUALITY

The quality of the milk produced by any herd of dairy cows is very important today—prices and profits depend on this. Milk quality is influenced by a number of things:

Breed All our breeds vary, with Channel Island breeds producing the highest quality and high-yielders often lowest.

Breeding Individual cows vary within a herd; the quality of their milk depends partly on their breeding. Breeding from bulls and cows with high quality milk usually helps to improve the quality.

Age The older the cow, the lower the butterfat, proteins and lactose.

Stage of lactation Butterfat, proteins and lactose are high at the start of a lactation, and fall off, until the last 4–6 weeks, when they start to rise again.

Disease Chronic mastitis lowers proteins and lactose in the milk.

Milking technique The strippings are richest in butterfat and the cow should be milked out properly.

Feeding Good feeding of dairy cows is essential, otherwise the quality of the milk will fall. Some fibre should be fed in the ration; if there is too little, the butterfat in the milk will fall. If the energy values of the ration are too low, proteins and lactose will fall.

Milk is a complete food—a mixture of fats, proteins, sugars, minerals, vitamins and other substances in water.

Farmers are paid for their milk on the basis of the weight of butterfat, protein and lactose in their milk. The weight of the milk constituents is determined by sampling and testing the milk each week, and monthly averages are calculated. The weight of each constituent is then multiplied by its value as derived from the market value of that constituent to give a price per litre each month.

Penalties are imposed on milk of poor hygienic quality or milk which contains antibiotics.

Milk Quantity

At the present time there is a danger of too much milk being produced in Europe; more than is needed for human consumption. For this reason, Milk Quotas have been introduced, which limit the quantities of milk which can be produced from any farm.

Milk Recording

This means keeping a record of the amount of milk produced by each cow in a herd. It can be done in different ways and various services are available, covering milk yield, milk quality and breeding standards.

The milk recording service, which was set up by the Milk Marketing Board, now operates under the name of National Milk Records. At least 5 different services are offered, fees being charged according to the level of service and the number of cows in the herd. You should get details of milk recording from the nearest office of National Milk Records.

The advantages of milk recording are:

- It helps you to know the milk production of your cows—both in quantity and in quality.
- Food given to milking cows is expensive, and has a great effect on their milk yield. Good records make it possible to feed them to the best advantage.
- Recording helps in planning a breeding programme and is essential in the management and sale of high quality cattle.

BEEF PRODUCTION

Beef is always an expensive meat, costly to produce, and there have often been times in farming when beef production has not been very profitable. In the past, many beef cattle were kept to use up the by-products of arable farming and to produce dung for the crops; this system is not normally found today. Instead, there is an increasing use of beef-type calves from the dairy herd (mainly Friesians) to produce intensive young beef. Beef cattle from pure breeding of the beef breeds and those cross-bred with dairy cows are also reared and fattened. Beef may be fed right through on the same farm, or reared on some farms and sold for finishing (for slaughter) on others. There are many different systems of beef production, and much movement of cattle from one part of the country to another, and within any local district.

We have some excellent beef cattle in this country, which produce high quality meat, but they have not been as fully tested, bred and standardised as pigs, bred purely for meat production. Many changes have taken place in beef production methods, with the introduction of new techniques, and more changes are coming.

Beef for the Modern Market

A good beef carcase must have:

- The right weight for its purpose
- Plenty of lean, tender meat; not too much fat or bone
- The greatest possible proportion of high-priced cuts

Some cuts in order of value:

Rump steak		(Obtain average
Sirloin		retail prices
Rib		per pound
Brisket		from your
Flank		local shop)

Seeing the value of the different parts of the carcase gives some idea of what to look for in the live animal. The best beef animals are well-fleshed (well covered with meat) with good development in the back and hindquarters, not too heavy boned and not too fat.

The beef animal can be judged as a whole, paying attention to the hindquarters (see the list of points on page 128).

While the ideal traditional beef beast, of the kind used to win all the fatstock shows, might be a small neat Aberdeen Angus bullock, this is not the type which suits modern needs best. The typical commercial beef animal of today is a good, young Friesian or crossbred bullock, lean, large-bodied and quick growing. Beef today is produced from bullocks and young bulls, and from heifers.

In practice, cattle vary so much—in type, rate of growth, and ability to turn food into meat—that often a batch will not all respond to the same treatment. If a farmer tries to produce early beef, some of the cattle may not finish in time, and will have to be kept on and fattened later.

The newer methods of early beef production depend on quick growth and no checks.

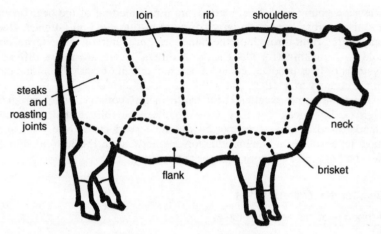

Figure 8　A carcase of beef

Calves for Beef Production

Most of our beef supplies come from calves produced by our dairy herds—either pure-bred such as Friesian, or crossbred with beef bulls on dairy cows. The price of calves has risen steeply in recent years, due to a strong demand for rearing calves from home and overseas. Because of the price of calves, farmers look for a good strong calf which will grow quickly and stay healthy.

Many of these calves are reared artificially and go into intensive beef units. Most dairy calves are born in the autumn, and if they are to be reared on grass after weaning, have the advantage of a summer at grass in their first year. Spring-born calves do not get much advantage from their first summer, as they are not able to eat much grass. Thus, with natural rearing, autumn-born calves are cheaper to rear than spring-born calves, and often grow more quickly.

Veal

Veal is the meat of a calf at about 3 months old, specially reared, grown quickly and fed on special foods. It should weigh 118 to 150 kg (260 to 330 lb) liveweight, with a deadweight 60 per cent of this (68 to 82 kg/ 150 to 180 lb).

There is little veal production in Great Britain, although there have been attempts to build up a veal industry. Little veal is eaten in Britain; it is much more popular in other European countries.

Veal can be produced by feeding whole milk but at present costs this is too expensive. Skim milk can be used, with added foods, but only if you can be sure of a regular supply. There are special foods available for calf feeding.

Beef

The killing-out percentage of beef animals (proportion of deadweight to liveweight) is 54–60 per cent, a low figure (compared with pigs) because of the large amount of stomach and gut in cattle.

Butchers prefer animals with a liveweight of no more than 455 kg (1,000 lb), giving a carcase of about 270 kg (600 lb) deadweight. Heifers carry more fat than bullocks and should be killed at lighter weights. Uncastrated bulls grow more quickly and produce leaner carcases.

The best beef animals should not be more than $2\frac{1}{2}$ years old, and many today are killed earlier—at 12 to 18 months; in intensive beef units, at less than 12 months.

Growth in cattle, as in other animals, follows a pattern with different parts of the body making growth at different stages (see also page 105).

- *Frame* The animal develops its bone structure, even in a 'store period' when it is making no other growth and is feeding on poorer quality foods.
- *Flesh* The animal fills in its frame with muscle and fat in a 'feeding period' when it gets plenty of good food.
- *Fat* In the later stages with older cattle, much more fat is put on the body and today this is not required.

The normal aim today is to produce beef cattle without having any store period by keeping them growing all the time—frame and flesh together—until they reach the right liveweight and are ready for slaughter.

Intensive beef means the production of a young, lean beef animal weighing no more than 450 kg liveweight at no more than 12 months of age, often less. There must be no store period, no checks in growth, and the animal must keep growing all the time. Cattle are housed and must be well fed and managed carefully. Cattle on this system can be upset easily by wrong feeding or inefficient handling; losses of valuable beef animals can easily upset the profitability of a unit.

Baby beef was the old name for the production of very young beef animals, and *barley beef* was the system which became widespread when cereal grains were cheaper and so were fed intensively to young cattle.

Semi-intensive beef also means the production of a young, lean beef animal, but to a heavier weight, and over a longer period. The cattle may be finished following a summer on grassland, or fed bulky food of good quality (such as hay, or maize silage) in yards, or have a combination of grazing and yard feeding. In any case, the system needs steady growth, no checks, and careful management.

Semi-intensive methods lead to the production of beef animals at $1\frac{1}{2}$ to 2 years old, weighing no more than 600 kg liveweight.

Traditional beef production 'Store cattle' are animals which have been reared on one or more farms, and then are sold—either to dealers or to other farmers. Many store cattle are produced from the grassland districts of this country and some are imported from Ireland. There is a lot of movement of store cattle about the country, and beef cattle may finish fat a long way from where they were born and reared.

In the past, many store cattle were brought from the North and West—from the grassland districts, into the arable areas of the country, and fattened to heavy weights in yards—fed on arable by-products of roots, hay and straw and some cereals. This system is found very seldom now.

Today, store cattle are bought for finishing on grass or in yards—well-grown animals of up to 2 years old, which are finished as cheaply and quickly as possible. The success of this system depends on good buying followed by good feeding and management, producing animals weighing up to 700 kg liveweight.

Heavier beef animals are not needed now for the domestic trade—meat for the ordinary household—but there is always a demand for catering (restaurants, hotels and canteens) and for processing (meat pies and other products).

PRACTICAL WORK

You should be able to do the following simple jobs with cattle: proper handling, haltering, know the signs of good health, take temperature and pulse rate, be able to judge weight, condition and age (by inspection of the teeth), have a good idea of the points of dairy and beef animals.

Plenty of practical experience and a certain amount of instruction will help you to take a Practical Proficiency Test or NVQ in this subject.

LIVESTOCK NUMBERS

Each year the Ministry of Agriculture publishes figures giving the numbers of various types of livestock. This information is collected from farmers each June. You should record these figures for the United Kingdom and also for your own county.

	United Kingdom	Your county	Your farm
Total cattle			
Cows and heifers in milk and in calf			
Bulls			
Total sheep			
Ewes for breeding			
Rams			
Total pigs			
Sows for breeding			
Boars			
Total poultry			
Fowls			
Ducks			
Geese			
Turkeys			

THINGS TO DO

1. Find out what the main breeds and crosses of cattle kept for dairy and for beef production are on local farms. Visit markets, local shows, and if possible, the Royal Show and the Royal Smithfield Show, to see as many cattle breeds as possible.
2. Study a local calf-rearing unit and see how the calves are kept, how they are handled and managed, and what foods are used.
3. Visit a good local dairy farm. Study the system of housing and milking, the milk handling and hygiene, and the feeding methods.
4. Visit a good local beef production unit. Study the system of production, housing, feeding and age and method of disposal of the cattle.
5. Visit a local slaughterhouse, processing plant or butcher. See good and poorer quality beef carcases, see the different cuts of meat, and check on local prices.

QUESTIONS

1. How would you tell the health and condition of the animals by handling and inspecting calves, and older cattle?
2. Which are the more important breeds and crosses used for milk production and for beef production now? Why these particular breeds?
3. What conditions are necessary to rear calves well, and what are the main causes of loss or of poor performance in calf rearing?
4. Why are the production of milk and its quality so carefully controlled and inspected? How is milk quality checked and tested?
5. Which parts of the beef carcase are most valuable, and what are the reasons for this? How can the quality of a beef carcase be controlled by the producer?

Chapter 2

Sheep

SHEEP ARE hardy, well-covered animals, usually kept in the open all the year round. Although they have been very much improved over the last 150 years, they are still not so intensively farmed as either pigs, poultry, or cattle.

They can be (and often are) kept under poor conditions, are managed and fed cheaply (mainly on grass), and they do good to the land and produce a reasonable profit. There are moves to make sheep production more intensive by breeding, management, and even by housing and special feeding.

SHEEP FOR MODERN CONDITIONS

The most important and the most profitable produce of British sheep is their lambs. Wool is secondary to this, producing only between 5 per cent and 10 per cent of total receipts per ewe each year. (In some other countries, wool is of first importance.)

Under poor hilly or mountainous conditions, it is not possible to produce fat lambs for slaughter, and there a flock of sheep will produce lambs which will be sold off to better land elsewhere, for further breeding or finished lamb production.

The meat of lambs (so-called until the end of the year in which they are born) and hoggets (into their second year) is sold in the shops as 'lamb'.

Mutton is the meat of older sheep, including ewes. There is little shop demand for it today, and it makes a lower price than lamb. Much of our mutton is used for manufacturing.

Like other animals, a good lamb must produce a carcase which has:

- The right weight for its purpose
- Plenty of lean, tender meat; not too much fat or bone
- The greatest possible proportion of high-priced cuts

Some cuts in order of value:

Leg		(Obtain average
Loin		retail prices
Shoulder		per pound
Neck		from your
Breast (belly)		local shop)
Head, lower legs, etc.		

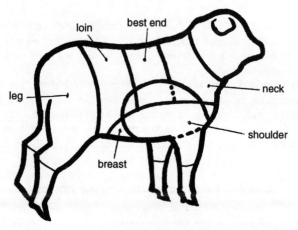

Figure 9   A carcase of lamb

Judging Sheep Quality

The points of the lamb carcase shown in Figure 9 give some idea of what is needed in the good quality live animal. We look for a good blocky body; well developed wide back, carrying plenty of meat; good development of the hind leg; not too much fat or bone.

Sheep are difficult to judge well: they are covered with wool and judging is by means of touch as well as by sight. It is good practice to study live sheep and place them in order of value, as well as to handle them.

Watch the following points in sheep which are for meat production: A solid blocky sheep, with short neck and deep wide back and body. *Back*—wide and well-fleshed. *Leg*—plump, well-fleshed, with meat well down to the hock. *Flesh*—firm and well developed (a 'wobbly' feel shows too much fat). *Skin*—bright and fresh looking. *Wool*—dense (not too open or lank).

A breeding ewe needs some of the qualities we look for in a lamb which will go for meat production. However, her first job is to produce the lambs, and milk well to feed them for quick growth. The ram will give to the lambs the high quality body type for meat production. *The ewe for milk and the ram for meat.*

Look for these points in breeding ewes: Long, low and level in shape. Not too fat at any time. Sound in mouth and udder, good teeth and two teats in good order. Good head, looking lively, and alert. Bright eyes and healthy fleece.

Condition scoring means rating sheep from 1 (lean) to 5 (fat) by handling. For best lambing results, ewes should have a score of about $3\frac{1}{2}$ when mated.

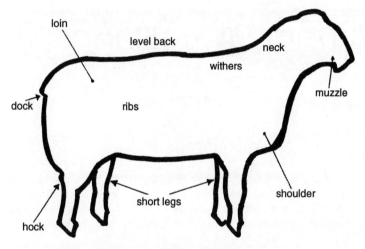

Figure 10   The points of a sheep

SIGNS OF HEALTH

There are some signs which show whether sheep are in good health and condition:

- Alert and active (a sheep must be able to walk well and move about comfortably). Bright eye, upright ear, good fleece condition, pink skin, not apart from the rest of the flock
- Temperature: average 39°C (102°F)
- Pulse rate (taken inside hind leg): 75–80 per minute

HOW TO TELL THE AGE OF SHEEP

There are 8 teeth in the lamb's lower jaw at birth. These are replaced by larger permanent teeth, from the centre first, as shown in Figure 11.

After this the teeth gradually get gappy, then fall out or get broken and worn down to the gums.

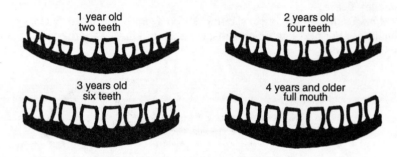

1 year old
two teeth

2 years old
four teeth

3 years old
six teeth

4 years and older
full mouth

Figure 11   Telling a sheep's age by its teeth

SHEEP BREEDS TODAY

There are more breeds of sheep (50 or more) in this country than any other type of livestock, and from these there are a vast number of crosses. The reason for this large number is that sheep, to do well, need to be suited to conditions of land and climate, and there is a lot of variation between conditions in different parts of Britain.

In practice, many breeds are quite local. It is usually wise to stick to breeds which have been found to suit local conditions.

Stratification is the movement of sheep from higher land to lower land as a definite policy. After ewes of the mountain breeds have produced 3 or 4 crops of lambs, these 'draft ewes' are moved down to lower land and mated with rams of the 'crossing breeds' to produce half-bred ewes. These are sold to farmers in lowland grass and arable areas for crossing with a meat-type ram (often called a 'terminal sire') to produce fat lamb for slaughter.

It is possible to divide our breeds into separate classes:

The mountain and moorland breeds are suited to the hardest, highest (and sometimes the wettest) conditions. They must be hardy and active, but due to poor conditions they are often slow growing and not very productive of lambs. The quality of their meat is good, although they do not produce a heavy lamb carcase. Their wool is usually long and somewhat coarse.

Scottish Blackface	Herdwick
Swaledale	Welsh Mountain
Rough Fell	Exmoor Horn
Lonk	Derbyshire Gritstone

The grass hill breeds are suited to lower and less hard conditions than the mountain breeds. They are active and quite hardy, and can produce lambs of good quality, and good wool.

Cheviot	Clun Forest
Kerry Hill	Beulah Speckled Face

The meat breeds (sometimes called the Downs or Shortwools) are kept mainly to produce rams for crossing with other breeds for meat production. They need good conditions, being early maturing (quick growing), with very good carcase quality, and excellent wool. Generally these breeds are meaty, blocky sheep. They are listed in order of size, the smallest first.

Southdown	Wiltshire Horn (Western)
Ryeland	Hampshire Down
Dorset Down	Suffolk
Shropshire	Oxford
Dorset Horn	

The longwoolled breeds are large sheep, suited to good conditions. They grow to a great size, producing plenty of rather coarse meat and a heavy growth of wool. In the past they were kept for producing mutton from folded arable crops or good grassland, and today are mainly found only in small numbers. Some of them have become rare breeds. They are listed in order of size, the smallest first:

Kent (Romney Marsh)	Wensleydale
Leicester	Devon & Cornwall Longwoolled
Border Leicester	Lincoln

There are also imported foreign breeds which have become established in Great Britain, such as the Texel, Charollais, Vendeen, Bleu de Maine and several others.

No notes are given here on the individual breeds of sheep. Fashions and market requirements change, and new breeds and crosses may be imported into a district or into a country. A booklet published by the

National Sheep Association under the title 'British Sheep' is interesting and useful to study.

For your own information, find the breeds and crosses of sheep which are important in your district, study them and take note both of the breeding ewes and of the rams which are used on them.

Cross-breeding

In spite of the large number of sheep breeds, cross-breeding is very common, to suit land and farming conditions and to produce lambs of the type needed. There are some 'crossing breeds' of which rams are mated with ewes of other breeds to produce breeding ewes. These include: the Border Leicester, Hexham Leicester (Blue-faced Leicester) and Teeswater.

There are several well-known regular crosses, which are commonly sold as cross-breds or half-breds under various names:

> *Scotch Half-Bred*: Border Leicester X Cheviot
> *Welsh Half-Bred*: Border Leicester X Welsh
> *Greyface*: Border Leicester X Scottish Blackface
> *Masham*: Teeswater X Swaledale
> *Mule*: Hexham Leicester X Swaledale or Scottish Blackface

BREEDING PRACTICE

As a rule, sheep are chosen to suit particular farming conditions. There are two points to consider:

1. Use the best sheep that the land will carry.
2. Buy the cheapest sheep you can, so long as the quality is right.

On the mountains, only hardy breeds will do. On the lower hills, pure lambs and cross-breds from the mountain breeds will thrive, and so also will the breeds specially suited to these conditions. In the lowlands, better sheep are needed to match the good conditions, and there it is common to buy in ewes from poorer conditions and to cross them with a meat-producing (Down) type of ram. Meat rams are commonly used on any pure or cross-bred ewes for fat lamb production, but the lambs produced are not usually kept for breeding. Meat rams may be used on any of the cross-breds listed above.

Choice of Ram

The ewe provides the milk and the means of rearing good lambs, while the ram provides the qualities that will produce a lamb with good quick growth and the right sort of carcase. It is very important, therefore, to select a type of ram which will do the right job.

A few years ago, a quick-growing large blocky ram, such as the Oxford Down, was suitable. Today, a fairly lean lamb is needed, which can be weaned and finished at a lighter weight, and for that reason the Suffolk is a very popular crossing ram. Fashions change, and market needs do not always stay the same, so it is important to keep up to date and use the right type of ram.

Ram Breeding

Pedigree and pure-bred rams for use of commercial sheep farmers (either for pure breeding or for crossing) are mostly produced from farms which specialise in this job. They rely on the higher prices from their rams to cover costs which are extra to those of normal sheep farming—more shepherding, high prices of bought-in breeding stock, special feeding and in some cases folding the sheep on special crops.

SHEEP PRODUCTION

We must remember that the chief product of a flock of sheep is their lambs. The ewe must be hardy and active enough to suit the conditions, productive (prolific), produce plenty of milk, and be a good mother.

A ram (male sheep) may be used for breeding carefully from 8–9 months old, but in his first year (a ram lamb) he must be used carefully.

A ewe lamb (female sheep in her first year) may be mated, especially if she was born early in the year and well grown. Although this is common practice now, most ewes are first mated in their second year.

Ewes come on heat only at certain times of the year (late summer and autumn) not all the year round like sows and cows. A ewe will come on heat every 16–17 days at this time until she is 'in-lamb' (the heat period lasts 24 hours on average).

The ewe is 'in-lamb' for about 5 months (21 weeks or 145 days). British breeds lamb once a year (with the exception of the Dorset Horn which will sometimes breed for 10 months out of the 12 and which can lamb twice in a year). The rate of production may be anything from 1–3 lambs per ewe, usually stated as a percentage—that is, the number of lambs per 100 ewes. 150 per cent is a good average, 200 per cent is the aim.

The ewe has only two teats and thus can rear two lambs; sometimes she can rear triplets, but if this is difficult it is necessary to 'foster' the extra lamb on to another ewe, or rear it artificially. Under intensive systems, lambs are taken from the ewes very early and reared on special milk substitutes.

It is usual to put 30 ewes to a ram lamb, 40–50 to an older ram. The rams are marked on the chest to show which ewes they serve, and put with a group of ewes; it is good practice to change rams around after 16 days, and if different rams are marked with different colours, this will show if the rams are working properly, and which ewes have not bred.

Flushing Ewes

After the lambs are weaned (often at 4 months), ewes can be put on poorer keep for a time during the late summer. Ideally they are then in good breeding condition; fat females do not breed well. A month or so before they are due to go to the ram, they should be fed well for breeding by being put onto better keep, such as a grass field which has been kept specially for them or a new ley (high quality pasture). This is known as 'flushing' and can help to produce more lambs.

In-lamb Ewes

During the first 3 months of pregnancy, the lamb is very small inside the ewe. During this time it is important to keep the ewe in good health and condition. Look after her feet, feed her reasonably well, keep her exercised and active, and control worms and liver fluke, if these are likely to cause trouble. Good feeding during the last 6–8 weeks before lambing is very important, as the unborn lamb grows fast and the ewe needs more food. Grass alone is not enough.

Lambing

Lambing is one of the busiest times of the year. It is best done in a sheltered field (not used for sheep for some time) or in new pens made of bales, or of hurdles and straw. It is important to worm ewes just before or after lambing.

Signs that lambing is near: the udder fills up and stiffens, the ewe gets uneasy and may go off on her own, she begins to strain, and a 'water bag' appears.

Most ewes lamb without difficulty but a few need help, Orphan lambs and the odd one of triplets may have to be put onto other ewes with singles.

Jobs which need to be done with young lambs include treating the navel with antiseptic to prevent infection (navel-ill), injecting against lamb dysentery, castrating the male lambs, and tailing (shortening the tail by rubber ring method or by cutting) to avoid soiling.

Although lambs going fat off their mothers are not castrated in some parts of the country, castration is still common practice. It can be done with the knife, by the bloodless (Burdizzo) method before one month old, or by rubber rings (Elastrator) in the first week of life.

Fat Lamb Production (for Slaughter)

A lamb weighs about 4.5 kg (10 lb) at birth if a single; twins will weigh 3 to 4 kg (7–8 lb) each. They rely entirely on the ewe's milk for the first two to three weeks, although they will start to nibble a little grass from 10 days old. It is important to have a type of ewe which milks well from grass without requiring other foods.

Lambs may be creep fed, either with food in a trough, fed where the mother cannot get at it, or by letting the lambs 'creep' through holes in a fence onto fresh short young grass, which, by having had a long rest since sheep were last on it, is 'clean' from sheep pests and other troubles.

Lambs may be ready to go for slaughter from $3\frac{1}{2}$ months old, although many are 4 months or more before they go.

Whether sold for slaughter or kept on, lambs are weaned at about 4 months, and the ewes put onto poorer grass to cut off their milk supply and to get them into lean breeding condition. It is important to keep an eye on the ewes' udders for a short time after weaning.

Store Lambs

Lambs not sold fat during the summer may be kept for sale or feeding as store lambs. They will be fattened off on grass or other green foods, such as rape, kale, or beet tops during the autumn or winter.

SYSTEMS OF SHEEP FARMING

Methods of keeping sheep vary widely in this country, according to conditions of land and climate. These are some of the chief systems:

Mountain and moorland   Hard, poor conditions allow only a low production, and there may be many losses. It is very important that the sheep are able to look after themselves—in many cases they must know their way about the land or they will come to grief. Lambing is late (April), and the crop of lambs is small (about 100 per cent). Typical breeds: Scottish Blackface, Welsh.

Grass hills   The land is poor but conditions are not so hard and unproductive as in the mountains. The sheep for this system must be active, hardy, and must be good mothers. Lambing will be timed for the lambs to meet the growth of new grass. There is a bigger crop of lambs than under mountain conditions (120–150 per cent). Typical breeds: Cheviot, Half-bred.

Grass sheep on lowland farms It is usual for sheep not to be the only livestock enterprise under this system. They may be run with beef cattle or with dairy cows, and in many cases will be just a sideline. It is important to have sheep good enough (and productive enough) for the conditions. This can be a profitable system, with a good crop of lambs (150–200 per cent), most of them sold fat off their mothers during the summer.

- Normal lambing time—February/March
- Early lambing—before the end of January

With very early lambing there is a higher return for the lambs, but costs are higher also, and very careful shepherding is needed.

Intensive grassland system Keeping sheep very closely together on grassland—as many as 6 or 8 ewes per acre (15–20 per hectare) with their lambs. It is necessary to creep feed the lambs, either by trough feeding dry food or by creep grazing. The latter system is being tried out by some farmers; it needs careful management and 'clean land' for the lambs (land which has been kept free from sheep for as long as possible) to avoid trouble from disease and parasites (chiefly worms) (see page 124).

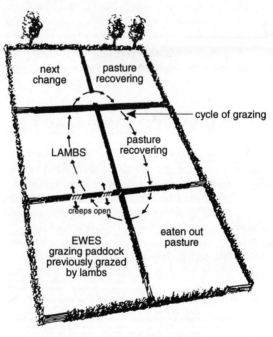

Figure 12   Forward creep grazing

Arable system The old methods of 'folding' sheep, closely fenced or hurdled on special green crops, are seldom found today. Instead sheep are often worked in with arable crops and grass, by the use of modern efficient electric fencing systems, labour-saving handling equipment, and other modern aids.

Systems of Sheep Meat Production

At the time of writing, there is a system of 'sheep quotas' which regulates the number of sheep which can be kept, or on which a subsidy can be claimed.

The MLC states that: 'the requirements of the home market and the export trade are becoming ever more precise'. For this reason, it is important to keep up to date with quality and grading standards for lambs, and to adapt the production system to sell meat animals at times when the prices are best. Study the weights, the grading standards, and the movements of prices to get the best returns.

Lamb carcase weights should be in the range 15–19 kg (33–42 lb). The grades are *standard*, 15–17.5 kg, and *medium*, 18–20.5 kg. Outside these weights there are light and heavy grades, which usually make lower prices. Carcases are also put into 6 classes; of these only Class 2 and 3L are good; the others make lower prices. Details can be found in MLC leaflets and price schedules.

It is important to produce and sell carcases in these grades and classes; otherwise the prices may be lower, and support prices may not apply— you may lose money. There is strong competition in Britain with imported New Zealand lamb, and the needs of the export market are very precise.

Mutton is the darker, more solid meat from older sheep, such as ewes finished for breeding. Little of it is sold in the shops, and most of it is used for manufacturing or processing.

Wool Production

The best quality wool is produced on the shoulders of the sheep. It is very important that sheep should be shorn well, to make the best use of the wool and to get the best price.

Under lowland conditions, shearing is usually at the end of May or early June, and is best in warm weather when the wool has 'risen' from the skin. It is important to shear under clean, dry conditions, avoiding double cuts, dirt, stains, dung, tar, straw, string, etc., all of which reduce the price.

The wool clip from the various types of sheep will average: longwool 4.5–5.4 kg (10–12 lb); lowland grass sheep 2.7–3.0 kg (6–7 lb); mountain sheep 1.8–2.25 kg (4–5 lb).

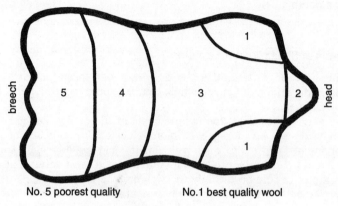

No. 5 poorest quality   No.1 best quality wool

Figure 13   Quality of the parts of a sheep's fleece

ROUTINE SHEEP CARE

In most flocks of sheep, foot troubles affect a number of animals. The commonest cause is the disease *foot rot*. It is possible to get rid of it by taking special care with treating affected sheep, by using the foot-bath regularly (containing a formalin or copper sulphate solution) by trimming overgrown feet with a knife or clippers, and by taking precautions not to introduce infected sheep into the flock. Sheep can be treated against foot rot by injection.

Dagging

Sheep and lambs, especially when grazing on rich pasture, scour (get diarrhoea) at times and become dirty round the hindquarters. This is also a sign that the animals are affected by worms. The dirty matted wool should be cut away using hand shears or a shearing machine). This job is important:

- before shearing, which is made much more easy and pleasant by doing this
- before ewes are put to the ram for mating

Dipping and Spraying

There have been periods when it was compulsory to dip sheep against the disease sheep scab, which is one of the notifiable diseases (see page 121), and also to control or avoid trouble from sheep fly (see page 123). Spraying has sometimes been permitted as an alternative method, and for fly control there are now chemicals which can be applied direct to the back and hindquarters of the sheep. Regulations about the form of disease and parasite control change from time to time; make sure that you know what is required by law, and that you keep good records of substances used.

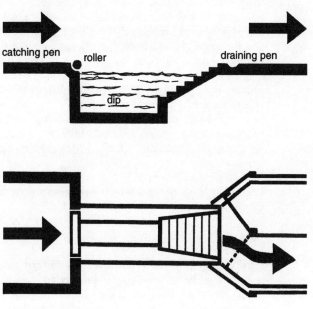

Figure 14   A sheep-dipping bath

Worming

All sheep carry some worms in their stomachs and intestines. Where many sheep are kept together, and the land is not rested from sheep, worms will cause trouble, particularly in lambs during the summer.

Good grazing management, and resting the land from sheep for as long as possible, will reduce or avoid worm trouble. Also, it is better for older sheep to follow lambs when grazing (creep grazing). (See Figure 12.) Worms are also partly controlled by dosing with various chemicals and by injection.

THE SHEPHERD'S CALENDAR

These dates and jobs apply to lowland grass sheep farming with lambing at a normal time of year. For other systems—such as very early lambing, or mountain sheep farming—the dates would be different, but the general principles would be much the same.

August	Select and condition score your breeding ewes (and ewe lambs); cull (get rid of) those which are worn out, old or barren and replace them. Check on health of feet, and the general condition of the flock. See that they do not have too much food, to get them in good breeding condition.
September	Flush the ewes, by putting them onto fresh keep or otherwise feeding them up. Dipping.
October	Turn the rams in to the ewes, according to the date when the lambs are wanted (5 months after mating).
November	Keep the flock grazing. Check on the feet and general condition.
December	Start to give a little extra food, and exercise the ewes.
January	Feed a little more, and watch health and general condition.
February	Feed the ewes to get them into good order. Watch general condition. Look out for twin-lamb disease. Prepare for lambing.
March	Lambing. A busy time when it pays to take great care of the flock. Extra feeding will be needed.
April	Take care of the lambs and keep them growing. Castration and tailing of lambs (if not done earlier). Creep feeding the lambs to get the fastest growth from them.
May	Make sure the flock has good fresh grazing. Start to dose or inject against worms. Dag the sheep (shear around their hindquarters) to prepare for shearing and to keep them free from fly attack.
June	Shearing, when the weather is warm and dry. Treat the feet of the flock, dose or inject again for worms if necessary. Start to sell fat lambs. Treat against fly.
July	Weaning the lambs. Drafting (selling out) unwanted ewes. Treat against fly.

PRACTICAL WORK

You should be able to do the following simple jobs with sheep. Proper handling, turning-up, weighing, know the signs of good health, take temperature and pulse rate, be able to judge age and weight, and have a good idea of the points of a good fat lamb and a good breeding ewe suitable for your district. Learn how to condition score sheep.

You should be able to handle lambs, mark and dose sheep and lambs, dip them, make sure they are making proper progress and recognise simple troubles.

Sheep shearing needs a good deal of practice, and it is important to start using a good shearing method. Modern style shearing, based on New Zealand methods, is fast, kind to the sheep, and avoids spoiling the fleece.

Plenty of practical experience and some instruction will help you to take a Practical Proficiency Test or NVQ in this subject.

THINGS TO DO

1. Visit local sheep farms with good methods of management and handling. See and study how the sheep are managed and handled, and find out the main points of sheep care in the summer and winter periods.
2. Visit a local slaughterhouse, processing plant or butcher's shop. Have a good look at both good quality and poorer quality carcases. See how they are cut up and study the various cuts of meat. Find out prices in the shop.
3. Visit local markets, sales and shows and see as many different sheep breeds and crosses as possible. If you get the chance, visit the Royal Show or the Royal Smithfield Show and see all the sheep, and sheep equipment, that you can. Visit a county agricultural show, or a demonstration arranged by the National Sheep Association (NSA).
4. Learn to handle sheep and lambs, and see as much as possible of routine care of a sheep flock at different times of the year, including lambing, weaning the lambs, care of the feet and sheep's health and condition generally.
5. Study the system of Condition Scoring sheep and learn the technique.
6. Study the MLC leaflet on Lamb Carcase production.
7. See good modern methods of sheep shearing (if you get the chance, have a go yourself), and of handling the wool. Find out weights of fleeces. See how sheep are dipped and otherwise protected against disease and parasites.

QUESTIONS

1. How would you tell the state of health and condition of the animals by inspecting and handling a pen of sheep?
2. What breeds and types of sheep are mainly used in your district, and how are they used for breeding purposes? What rams are used and how many of them to a flock of sheep?
3. What are the main reasons causing loss of lambs, or poor growth and performance of sheep, and how can these be avoided? How do you aim to keep sheep healthy?
4. With the type of sheep found locally, what is the weight, condition and age of a good fat lamb? How can you measure or judge this?
5. With the type of sheep found locally, what weight of fleece would you expect to get? What can be done to get the best possible price for the wool?

Chapter 3

Pigs

THE PIG has no coat and will not thrive under cold conditions. When kept properly it is quick growing and efficient at turning its food into meat. Today many pig units are intensive with animals farmed on a small concreted site, although outdoor pig breeding has increased. Modern pigs have been carefully bred and selected; they need high quality management and controlled conditions to give their best performance. In good modern pig farming nothing is haphazard or left to chance.

PIGS FOR MODERN CONDITIONS

Good quality pigs are needed for three purposes:

1. Pork (fresh pigmeat)
2. Bacon (pigmeat which is factory processed and cured, after grading to very strict standards)
3. Manufacturing (pigs are processed in the factory to provide pork, bacon, hams, pies, sausages, tinned meats and other products)

For both pork and bacon the same type of long, lean pig is needed. For manufacturing, a broader and heavier type may be suitable.

A good pig which is to be sold for any these purposes must produce a carcase which has:

- The right weight for its purpose (see pages 49 and 50)
- Plenty of lean meat; not too much fat or bone
- The greatest possible proportion of high-priced cuts

    Some cuts in order of value:

Gammon (ham)
Back
Fore-end
Streaky (belly)
Head, hocks, etc.

(Obtain average retail prices per pound from your local shop)

43

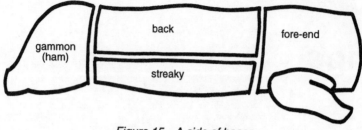

Figure 15   A side of bacon

Judging Pig Quality

What we have considered already gives some idea of what is needed in
the good quality pork and bacon carcase. Some of these things can be seen
in the live pigs, where we look for: Long body. Leanness. Light head and
fore-end. Good ham development. Fair depth of body. Fine bone.

A famous Suffolk farmer, the late Mr David Black, said that a good
sow should have 'the shoulders of a princess and the bottom of a cook'.
It is not easy to judge the live pig and the only real test is the carcase
which it produces. However, it is good practice to look at live pigs,
compare them, and watch the following points:

Back—long, level wide. *Side*—long, deep, smooth, of an even depth.
Gammon—broad, full, but not flabby. A straight underline with at
least 12 well-placed teat spots. *Shoulders*—light, level, medium width.
Head—light, with a light jowl. *Flesh*—firm, without too much fat.

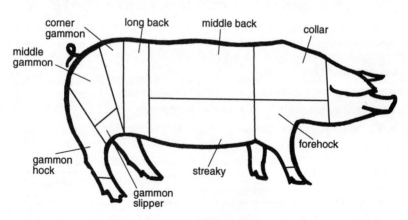

Figure 16   Bacon cuts

Skin—fine, thin, free from wrinkles. *Tail*—a thick root (not thin and rat-like).

Ultrasonic testing   During recent years, a measurement of the depth of fat and meat in the live pig has become possible by using an ultrasonic tester. This is an electronic depth-sounder which gives a very accurate estimation. This practice allows selection of replacement breeding stock.

SIGNS OF HEALTH

These are some signs which show whether pigs are in good health and condition:

- Vigorous, alert (they look at you when you come in to them), springy tail, moist snout, warm ears, skin in good condition (not scaly or discoloured). Good appetite, firm dung. Quiet, steady breathing, no coughing
- A normal pig barks when disturbed, stretches when it gets up, eats well, and does not stay away from other pigs
- Temperature: average 39°C (102°F)
- Pulse rate (taken inside hind knee or over the heart): 70–80 per minute

PIG BREEDS TODAY

During recent years there have been rapid changes in our pig types and breeds, because of the needs of the meat trade and the farmer's need to keep pigs profitably. Several of the old breeds have declined in numbers and even disappeared. Other breeds have been imported, and there has been a great development of hybrid pigs, sometimes bred from several different breeds and strains of pigs.

Types

The modern *quality type* of pig, formerly called the 'bacon type', but suited to all modern purposes, is long, lean and fast growing. The main breeds are Large White, Landrace and Welsh.

There are also several hybrid types, the products of commercial breeding companies and group breeding systems.

The *coloured breeds* have lost popularity now and are few in numbers. They have deep, rather short bodies, are quick growing but get fat rather early. The sows used to have a reputation as good mothers. The main breeds are British Saddleback and Large Black.

The more recently *imported breeds* have come into this country for special purposes such as scientific breeding programmes. These breeds include Hampshire and Duroc (from America) and Pietrain (from Belgium).

The *rare breeds* are kept in small numbers, often by enthusiasts, sometimes as breeds of local interest. They include the Tamworth, Middle White, Berkshire, Gloucester Old Spot and Large Black.

Breeds

Large White   Known in some countries as the *Yorkshire*. A white, long, quality type pig with upright ears; well suited to intensive conditions. The breed has been much improved to meet modern needs. It started in Yorkshire, but has spread through Britain and is found in most parts of the world. This is now the standard British breed.

Landrace   A white, long, lean type of pig, with lop ears and long thin head and neck. Our type of Landrace was first imported from Sweden in about 1950. It is an important breed, with the advantages of great length and a light head, which has proved excellent for bacon production.

Welsh   Another important white, long, bacon-type pig, with lop ears. It has been called 'the British Landrace'. In the old days, there was much variation between different strains of Welsh pig, but today by selective breeding it has become a much more standard type.

British Saddleback   A black pig, with lop ears and a white band over the shoulders and front legs. The breed resulted from the amalgamation of the Wessex and the Essex Saddleback breeds. It produces a sow with good mothering qualities, which can do well outdoors. It is sometimes crossed with a white boar to produce a 'blue pig'.

Hampshire   A breed imported from North America. Prick-eared, with a white saddle on a black body. Thick set, quick growing, and used for cross-breeding to produce a pork pig for fresh meat production.

Duroc   Imported from North America. Lop-eared, ginger brown in colour, used for cross-breeding; has a very quick growth rate.

Pietrain   Imported from Belgium. Prick-eared, with irregular dark spots on a white body. Good muscle development, pale coloured meat. Used for cross-breeding, producing a very lean carcase.

USE OF THE PIG BREEDS

In pig farming, the common practice is to use a white, quality-type pig which may be crossed in various ways to produce good quality commercial pigs. Today, most commercial units use hybrid breeding sows.

The farmer aims for big litters and good mothering by the sows, quick growth, food economy and carcase quality.

Cross-bred pigs from parents of different breeds or different strains are often stronger and 'better doers' than pure-bred pigs; this quality is called 'hybrid vigour', and it can lead to better growth rate and feed conversion.

Examples of Common Breeding Practice

- Large White, Landrace, or Welsh—pure-bred
- Landrace or Welsh crossed with Large White
- Large White crossed with cross-bred (hybrid) sows
- There are many commercial crosses of pigs—produced by companies which specialise in pig breeding—marketed under trade names. They are often sold to other countries.

BREEDING PRACTICE

A boar (male pig) may be used for breeding from 8 to 9 months of age. A gilt (young female pig) may be mated from 200 days old, when she is well enough developed (about 100 kg, 220 lb, liveweight) although some people would delay this until she is 8 months old. She will come on heat every 21 days until she is 'in pig'—the heat period lasts 2 to 3 days. The first heat after the birth of a litter usually comes 3 to 4 days after the litter is weaned but this will depend on the suckling period. Sows weaned earlier than 3 weeks may not come into season for 10–14 days after weaning.

The sow (female pig after she has had a first litter) or the gilt is 'in pig' for about 115 days (3 months, 3 weeks, 3 days) and commonly suckles her litter for about 3 to 5 weeks. She is then mated again as soon as possible. Very early weaning, before 14 days, is sometimes practised.

Early-weaning systems are designed to get the sow to farrow (produce a litter) as often as possible, in order to get maximum production from her. The aim is to produce 22 pigs reared per sow per year, and these figures show how it can be done:

Litters per year	Pigs reared per litter	Pigs reared per year
2.5	8.8	22
2.4	9.2	22
2.3	9.6	22
2.2	10.0	22

During her lifetime, a sow may produce a total of 6 to 7 litters, although an average in practice is more like 5 to 6 litters.

Artificial insemination (AI) is used with pigs as an alternative to natural mating, and it can be quite successful; only semen from top quality boars is used.

Signs of Coming on Heat

- General behaviour different from normal
- She may go off her food
- She may try to break out of building or pen
- Vulva red and inflamed
- Riding by other sows

When a sow or gilt is fully on heat (ready to take the boar) she will stand still if her back is pressed down or if the pigman sits astride her. Sows due to come in season, or being checked for return to service, should be kept within sight, sound and smell of the boar.

One boar will be needed for up to 18 female pigs; in outdoor systems, 1 boar will be needed for 12/14 sows. A young boar should not be overworked at first. It is best to allow mating on the second day of the heat period and to allow two services, as this often means larger litters.

Sows and gilts should be well fed, but not overfed, during the time they are 'in pig', to make sure of litters as large and well nourished as possible. Good quiet, careful management is needed; stress causes problems (see page 109).

Signs of Farrowing

- Sow or gilt tries to make a nest with bedding
- She becomes restless
- Milk present in teats (usually within 12 hours of farrowing)
- Enlargement and swelling of vulva
- Slackening of muscles either side of tail

THE LITTER

Careful feeding, breeding and management will help to produce a large litter. It is very important to rear as many piglets as possible, but there are a number of risks and it is common to lose 12 per cent of all pigs born before weaning. Quiet and careful handling of sows and gilts will help to prevent nervousness; if disturbed, sows often attack their piglets. A farrowing crate or rail in the pen allows the piglets to escape from their mother during the first few days; there is a danger that she may lie on some of them. Infra-red lamps are commonly used to warm the young piglets and to attract them away from their mother into a special creep area. Farrowing areas should be clean, dry draught-free and warm.

Most sows are farrowed indoors, and young pigs reared under indoor conditions may suffer from piglet anaemia (shown by scouring, weakness, pale colour), unless iron is given to them by injection or by dosing during the first week of life.

The sharp eye teeth of the little pigs are usually removed by nipping off with pliers, to avoid irritating the sow's teats.

Young pigs grow better if given food as well as their mother's milk; this 'creep food', which is fed where the sow cannot get at it, starts at 7–10 days of age onwards and is made specially tasty and digestible. The piglet at birth weighs about 1.3 kg ($2\frac{3}{4}$ lb) on average. A poor birth weight may be a sign of poor feeding of the sow, a breeding weakness or old age. In pig recording, piglets are commonly weighed at weaning. This is the only way of 'milk recording' the sow—because the piglets have fed mostly on their mother's milk for these first weeks of life. An average weight at 3 weeks is 6 kg (13 lb) per pig. (After 3 weeks the sow's milk is reduced in quantity, and it is essential that the pigs have plenty of good quality food.)

Weaning is usual at 21–28 days. Aim for a high weaning weight, which usually means quicker growth later, and better quality pigs. An average 8-week weight is 18 kg (40 lb), but aim for 22 kg (48 lb).

PIG PRODUCTION

The income from a herd of pigs comes from the sale of pigs for meat or for breeding; there is no other source of income, such as milk or wool. For this reason, the quality of the pigs produced is of great importance and the costs of production need very careful attention. The following are the different possible types of production from a herd. More than one may be practised at the same time on any farm.

Breeding stock Anyone looking for breeding stock to start or to increase a breeding herd will want animals of good type with good records of performance. This type of pig farming is a very specialised business which needs first-class management and up-to-date thinking.

Young pigs Some farmers breed only and sell the young pigs as weaners (5 to 8 weeks old) or stores (10 to 12 weeks). A good type of pig will make the best price; often these pigs are sold by weight—so much per kg liveweight—and this is one main reason why a high weaning weight is important. These pigs are bought by feed firms or by groups of farmers, or are sold direct from one farmer to another, who finishes them.

Pork/cutter Pigmeat sold fresh. There is a bigger demand for pork in the winter than in the summer, and prices rise around Christmas time. Local standards vary, but suitable pigs are generally lighter than the bacon weights. Lightweight porkers are about 60 kg (130 lb) and heavy porkers, or cutters, 80 kg (175 lb) liveweight. An average pork weight is 70 kg (155 lb). Quality is important today, and what is needed is a well-grown, well-fleshed pig—lean, with fine bone and plenty of meat in the right places (see page 44). Nearly all our pork supplies are home produced.

Bacon Pigmeat cured in the factory, and sold either 'green' or smoked. A lean pig of the highest quality is needed; the price varies according to quality, and grading is very strict. An average 90 kg (200 lb) liveweight. Nearly half our bacon supplies are home produced.

Which Type of Production?

The first decision for the pig farmer is whether to breed only, fatten only or breed his own pigs and fatten them. Land, buildings, capital available, local sources of weaners or store pigs—all these must be considered. Some people are more successful with either breeding or feeding—these are both specialist jobs. Britain produces much of its own pork, but a large proportion of our bacon is imported (the percentage varies from time to time).

There is a lot of discussion about pig production generally. Bacon production is profitable if you have a good lean type of pig which grades well and consistently. If the pigs are not quite up to this standard, pork production will be less troublesome, but may earn less profit. Heavy pig production (for manufacturing) needs a pig which must be kept longer but may be fed more cheaply.

The porker and the baconer must be produced, on the farm, to suit very strict needs. Some pigs can be 'tailored' by the factory to meet their standards, and thus can be produced on the farm with less attention to detail.

Grading

The grading of all types of pigs for meat production is most important, particularly for bacon, as it affects the price to the producer. Grading the live pig on its appearance is not reliable; the grading of the carcase after slaughter is much more reliable. A bacon carcase is graded to one of four standards. A first-grade carcase would be 60–75 kg deadweight. Details can be found in MLC leaflets and prize schedules.

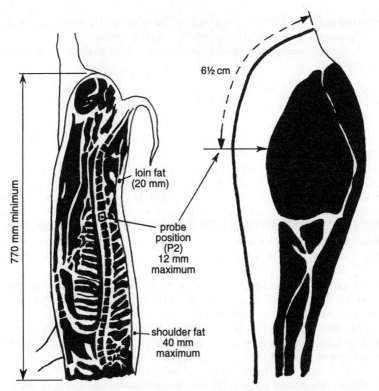

Figure 17   Assessing pig carcase quality

Killing-out Percentage

This shows the amount of usable carcase obtained from a certain liveweight of animal. It is high in pigs compared with other animals. For example, a live baconer of 90 kg (200 lb) will give 68 kg (150 lb) deadweight. About 14 kg of this will be head, legs, etc., and the remaining 54 kg will divide into two 27 kg sides.

	Average liveweight	Deadweight	Killing-out per cent
Pork	70 kg	50 kg	70
Bacon	90 kg	68 kg	75

Growth of the Pig

Pigs are quick-growing animals and need high quality foods, good housing and careful management. Modern pigs need carefully controlled conditions (warmth, ventilation, etc.) to do their best.

Like other animals there are three stages in a pig's growth, in this order:

- Bones, head and insides
- Muscle (meat)
- Fat

The shorter, deeper, quick-growing type of pig (the old 'pork type') starts to lay on fat at about 45 kg (100 lb) liveweight. The longer, leaner, quality 'bacon type' does not put on much fat until it is 55 kg (120 lb) liveweight. Gilts are leaner than hog (castrated male) pigs. Boar pigs have traditionally been castrated, but in recent years there has been a move away from this practice. Pigs are left *entire* (uncastrated) for meat production; they grow faster and there is no taint (strong flavour) in the meat until they are at much greater weights than is normal for slaughtering. The rights and wrongs of this development are still being discussed.

As the pig grows older, it puts on more liveweight each day, but needs more food to do this. Also, as it grows older more of the food is turned into fat rather than into lean meat.

For these reasons, it is better to feed the pig well when it is younger, and to restrict the amount of food when the pig is older. Because of better breeding stock, it is now possible to feed pork ad lib to slaughter weight (above 60 kg liveweight) without forming excessive fat.

Food conversion ratio (FCR) is a measure of the amount of food needed to produce a certain amount of liveweight on the pig. On this (and on the quality of the pig) depends the profit or loss made by the fattening pig.

A food conversion ratio at about bacon weight (100–110 kg liveweight) might be 3.0. This means that the pig needs 3 kg of food to produce 1 kg of liveweight. For examples of this see page 88.

Speed of Growth

The faster the growth of the pig (within reason), the better it is. We should aim to produce a 90 kg baconer in 120–130 days and a cutter in 90–100 days.

Management of Fattening Pigs

Most fattening pigs today are kept under artificial conditions—indoors, well-insulated, warm, controlled ventilation, very careful feeding, watering and management. Important points in management are:

Warmth   The more comfortable the pigs, the better they will grow and use their food.

Dryness   Wet causes discomfort and trouble.

Ample trough room   Plenty of room to eat.

Separation into batches according to size. This is to avoid fighting and bullying, and also because pigs of one size need the same treatment. Pigs are also separated now according to sex—hog pigs and gilts separately—because they grow at different rates.

Cleanliness   Pigs like to keep clean and should dung away from their bedding and feeding areas. Freedom from parasites, either inside the body, like stomach worms, or outside, like lice.

Avoid fright, sudden changes of management or food, anything which will check their growth.

Correct weight is important, for feeding, for sending away for slaughter, and for recording growth, and this can only be found out by weighing in a proper weigher.

PIG HOUSING

Pig housing is a subject which arouses strong feelings. Animal welfare enthusiasts have much to say about it.

Pigs can be kept successfully under all sorts of conditions; systems of housing and management come and go; fashions operate in such technical matters as housing and feeding pigs and influence many people. It has been said that there are as many designs of pig buildings as there are pig farmers.

First, we should remember what the pig needs: comfort, which means protection from the weather, damp, draughts and chills. Also fresh air, an opportunity to keep warm and clean and healthy surroundings.

What the pig farmer, and the pigman, needs are good food conversion (using as little food as possible to produce pigmeat); good health of the stock; an easily run and labour-saving system.

Modern types of fattening houses take all these needs into account, and breeding stock are also often housed in a similar way. We can still find pigs housed in makeshift buildings, in yards and even in the open air, but most pig units of any size today have become more intensive, and many pigs spend all their lives inside buildings. A major choice of system is between straw bedding and a slat/slurry system.

Ideal Conditions for Pigs

Much research work has been done in recent years on the best conditions, especially for fattening pigs; there are still different opinions about many points, but general agreement on several, such as:

Temperature   Pigs use part of their food to produce warmth, and at different ages they seem to do best at certain temperatures:

Piglets during first few days of life	21–26°C (70–90°F)
Piglets up to 4 weeks old	18–21°C (65–70°F)
Sows, boars and fattening pigs	12–18°C (55–65°F)

Temperatures should be taken near floor level, where the pigs live. High temperatures are not necessary throughout a building, but only where they will most affect the pigs, which should have a chance to escape the heat if they want to.

Insulation is most important to keep warmth in a building. Most heat is lost through the roof (and through too much ventilation). Insulation of the floor is important, to avoid cold and damp affecting the pigs.

Ventilation should keep the air fresh, healthy, clean and dry, without draughts. In modern houses it is controlled.

Drainage must be good, to avoid dampness. Pigs should dung away from the living and sleeping space. Some buildings have slatted floors in the dunging passages.

Trough space should be 30–40 cm (12–15 in) per pig. If there is too little this can cause trouble and weaker pigs may not get enough food. On some farms, food is given on the floor; this can work well if the floor is kept clean. With ad lib feeding, allow 10–15 cm (4–6 in) per pig.

Welfare   The official Welfare Code for livestock in this country gives standards and dimensions for different classes of pigs. This should be studied in detail.

Breeding stock are mainly housed indoors today, in specialist buildings for ease of work and management. In the past it was common to house breeding pigs simply and run them outdoors. This can work so long as they are protected from bad weather and can sleep warm and dry; it is a system occasionally found on light, well-drained soil, but it is not common.

Farrowing sows and their litters need careful housing. The piglets need warmth, with a farrowing crate (or sometimes a rail) to protect them from their mother, and a special creep area for warmth and separate feeding. After weaning, the piglets are moved to grower accommodation and the sows back into the dry sow accommodation for re-mating.

There are various methods for housing dry sows, and several of them are widely found in practice. Sows can be kept in yards, using straw for litter; sometimes kennels are constructed to provide warmth and comfort. It is important to have individual sow feeders, so they can be fed the right amount of food, without fighting and stealing.

Another method is to keep the sows in cubicles, where they lie separately but can come together in a yard.

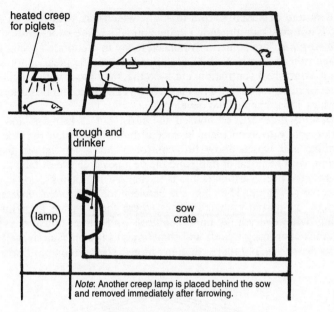

heated creep for piglets

trough and drinker

lamp

sow crate

Note: Another creep lamp is placed behind the sow and removed immediately after farrowing.

Figure 18   Farrowing unit

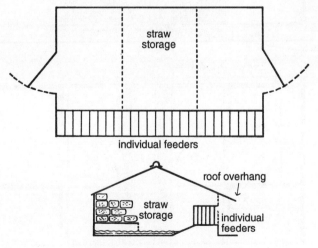

straw storage

individual feeders

roof overhang

straw storage

individual feeders

Figure 19   Covered sow yard

The system of keeping sows in stalls, which has been used in recent years, is not now recommended in the animal welfare code.

Growing pigs are sometimes housed today in elaborate special buildings and other structures which are aimed to provide ideal conditions—warmth, controlled ventilation, etc.—so that they grow without check. A number of firms make buildings just for this purpose, and it is interesting to see how ideas change and develop.

Fattening pigs can be housed in yards, but it is more and more common today to keep them in special buildings which provide ideal conditions and which allow full control of heat, ventilation, feeding, watering, with disposal of dung and the best possible use of labour.

There have been various fashions in the housing of pigs, and systems have come and gone. The choice is between the *controlled environment* buildings, on partially slatted floors where the pigs are kept inside a building with everything under control, and other housing systems which make part use of yards and sometimes of the open air as well. Two systems are shown but there are many more to be seen on pig farms.

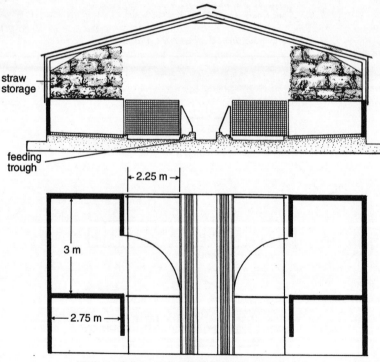

Figure 20   Accommodation for fattening pigs

Outdoor pig production   This system, which was practised a little in England in the past, sometimes attracts greater interest, which is the case at the time of writing.

It requires well-drained light land, with some protection from wind and weather. Some people are very successful with this system. Usually, the breeding sows are housed in small huts or arks. They produce their litters, which are run in the open for a time, and then brought into buildings to be finished.

PIG RECORDING

In recent years, the use of computers has had a great effect on pig recording.

This can be done privately, through commercial schemes or through the official national scheme run by the Meat and Livestock Commission (MLC). Recording shows the strengths and weaknesses in the system and provides forward planning so that profit can be maximised. For breeding stock, good basic performance records are essential to provide a basis for culling and to enable the herd to be run at maximum strength at all times.

Running a profitable pig business is not easy. For this reason it is important that staff should be technically competent and should keep abreast of new developments. The following table taken from MLC data will give an idea of current standards:

	Average	*Good*
Pigs reared per sow per year 19–25 day weaning	21.6	25.0
Age at slaughter		
60 kg liveweight	18 weeks	17 weeks
70 kg liveweight	20 weeks	19 weeks
80 kg liveweight	23 weeks	22 weeks
90 kg liveweight	24 weeks	22 weeks
Food conversion rate (FCR)		
to 60 kg	2.3	2.0
to 90 kg	2.5	2.4

Weights at different ages give a good indication of pig performance, and these figures are targets in a good unit:

Age	kg	lb
1 week	2.5	$5\frac{1}{2}$
2 weeks	4.0	$8\frac{3}{4}$
3 weeks	5.5	$12\frac{1}{4}$
4 weeks	7.5	$16\frac{1}{2}$
5 weeks	10.0	22
6 weeks	13.2	29
7 weeks	16.5	$36\frac{1}{4}$
8 weeks	20.0	44
10 weeks	27.4	$60\frac{1}{2}$
12 weeks	36.0	$79\frac{1}{4}$

PRACTICAL WORK

You should be able to do these simple jobs with pigs: proper handling, weighing, know the signs of good health, take temperature and pulse rate, be able to judge age and weight, have a good idea of the points of a good pig.

You should be able to handle young pigs, remove their teeth, dose or inject with iron, make sure they are making proper progress and recognise simple troubles.

Plenty of practical experience and a certain amount of instruction will help you to take a Practical Proficiency Test or NVQ in this subject.

THINGS TO DO

1. Visit local intensive pig farms with good methods (taking care to observe all the necessary health precautions). See and study the methods of breeding, rearing and feeding the pigs.
2. Visit a local slaughterhouse, processing plant or a good butcher's shop. Look at both good quality and poorer quality carcases, see sides of bacon or pork cut up, and study the various cuts of meat.
3. Find out from a local supermarket or butcher's shop the current prices for the different cuts of pork and bacon, and compare these with the price paid to the farmer. When you eat meat, see if you know which part of the animal it comes from.
4. Find out on local farms such facts as average litter sizes, litter

weights, individual weights of older pigs, losses of pigs and the reasons for them. Find out the prices paid for weaners from local farms.

5. In buildings used for pigs on local farms, study the details of construction, space for pigs, arrangements for feeding and watering, methods of dung disposal, and methods of insulation.

QUESTIONS

1. How would you tell the stage of health and condition of the animals by inspecting a pen of pigs?
2. What breeds and types of pigs are mainly used in your district, and how are they used for breeding purposes? At what age are pigs first mated?
3. What are the main causes of pig losses in their early days and how can these losses be avoided?
4. What are the main points to consider in housing, breeding and fattening pigs to get the best performance from them?
5. What grading standards are used now to decide the quality of a pig carcase and the payment for it? What performance standards do we expect from a good pig unit—numbers born, numbers sold, growth, feed conversion?

Chapter 4

Feeding Livestock

THE FEEDING of farm livestock is in many ways the most important part of animal management. It is completely within the control of the farmer and the stockman. It has a tremendous effect on the performance and production of the animal, on its health, its breeding, its well-being. And in the long run, proper feeding will affect the profit or loss made from the animal. Food is the largest single item of cost in animal production—as high as 80 per cent of pig and poultry costs, 60 per cent of cattle costs, less for sheep but still 50 per cent or more.

WHY THE ANIMAL NEEDS FOOD

In a few words, the animal needs food to keep it alive and to allow it to do its job. We must look at these needs in more detail.

Food for Maintenance

The animal needs food to keep it alive, to keep it in good health and good condition, and to keep it warm (to produce body heat). All these needs are for energy (or heat, which is a form of energy).

Also, as the animal body lives and does its work, some parts of that body are breaking down and so need repair and replacement. It needs some of the construction and repair foods (proteins).

So it is clear enough that for maintenance (and repair) the main needs are for energy foods, with a little protein.

Food for Production

No animal is kept just to live and stay in the same condition, unless it is a pet. It is kept for a purpose—and in most cases that purpose is production. There are several different types of production, and any one animal may at times be expected to do several of them at once.

Growth   Young stock are growing into full-grown animals. Meat-producing animals are putting on muscle and fat.

Work   Horses and sheep-dogs are kept for work. Oxen were

previously kept for this purpose in Europe and still are in some countries.

Milk Cows are kept for milk production. Any breeding female (sow, ewe or mare) produces milk for its own young, and this can sometimes be used for other purposes, such as cheese production.

Wool In this country sheep are kept mainly for meat production but wool is still an important product.

Young All breeding animals produce young—cows and ewes once a year, sows twice a year. Poultry produce eggs, and although most of them are not kept for hatching it is just the same process.

Food is needed by the animal for all these forms of production—the greater the production, the more food is needed. Some of the need is for energy foods, but there is also a great need for protein—for the growth of muscle, for the young growing within the mother, for the protein in milk, and for wool and hair. So while production foods must contain plenty of energy, they must contain a larger proportion of protein than the animal needs just for its maintenance.

What Type of Food Does the Animal Need?

It is obvious that not all animals need the same foods: it would not be right to feed a sow on hay, a heavy-milking Jersey cow on barley straw or a dry ewe on fishmeal. We must know what foods to feed, and why they are suitable or not suitable for a particular type of livestock. These are the things to consider:

The type of digestive system Pigs and poultry cannot deal with much bulk or much fibre. They need such foods as cereals and concentrates. Calves and lambs in their early days can only take a small amount of fibre, and so need digestible and fairly concentrated food, with milk and good quality grass. (This is discussed in more detail on pages 71–77.)

Age and size Young stock need food for rapid growth as well as for warmth and energy. Their food must be easily digestible and it must be palatable.

What the animal is producing This covers the production needs of the animal. It is either growing, fattening, milking, working or pregnant—and needs feeding accordingly.

Food Constituents—What Foods Contain

To understand the use and the value of farm foods, we must get to know what foods contain, and how these food constituents affect the animal.

All farm foods contain various proportions of these:

Water is present in all food. There is a high water content in roots and green foods; a low moisture content in most concentrates and fodders.

Water is needed in digestion and to replace moisture lost by the body. The body needs it for many purposes—to transport foods and waste products about, to remove waste products, etc.

Energy constituents are made of carbon, hydrogen, and oxygen (CHO). They provide the body with fuel from which is produces energy (work), and heat. Anything not used in this way is stored, mostly in the form of body fat. The main energy foods are:

- Sugars and starches, which are carbohydrates.
- Oils and fats, which are more concentrated, having about $2\frac{1}{2}$ times the energy value of the same weight of sugars or starches. Oils are liquid and fats are solid at normal temperatures.
- Fibre is not easily digested. It provides bulk in the digestive system, and when broken down provides energy.

Protein constituents are made of carbon, hydrogen and oxygen plus nitrogen (CHO+N). As well as providing some energy they also supply the body's needs for growth and repair. There are animal proteins (which in some ways have a higher feeding value) and vegetable proteins. Proteins and similar substances are made up of amino-acids; these include lysine, cystine, histidine, leucine, tryptophane and others.

Minerals for Livestock

Only small amounts of minerals are needed by the animal body, but a shortage of one or more of them can cause many problems. Troubles can be expected if no minerals (or not enough of them) are being fed in the ration. Most of the mineral matter in the body is in the bones and teeth—the rest is in the blood and tissues, where it is responsible for such important functions as the proper working of the digestive system.

The main minerals are calcium, phosphorus, sodium, chlorine, magnesium, sulphur, potassium, iron. Several of these are supplied naturally in a normal diet. We are mainly concerned with those which are needed by the body for the following purposes:

growth of bones and teeth—calcium, phosphorus
blood and digestive system—sodium, chlorine, iron
milk production—calcium, phosphorus, chlorine

The trace minerals or *trace elements* are copper, cobalt, manganese, also zinc, fluorine, arsenic, molybdenum. These are only needed in very small quantities.

How to Feed Minerals

There are two ways of getting livestock to take the minerals they need:

1. Mineral licks are given so that the animals help themselves—the idea being that they know what is good for them and take enough. Rock salt is often put out in the field, and this supplies sodium and chlorine (salt is sodium chloride), and specially prepared mineral licks and blocks are often put in the field or in yards and buildings.
2. A more definite method is to add mineral mixtures to rations being fed. This can be done practically only when a concentrate mixture is being given. Several firms make mineral mixtures which are specially designed for different classes of stock. A simple home-made mineral mixture can be made containing feeding chalk or limestone, steamed bone flour and salt.

Mineral Needs

A deficiency (shortage) of any mineral can cause problems. These are examples of mineral needs:

- Young animals need plenty of calcium and phosphorus for bone and tooth formation.
- Pregnant animals need both of these minerals for bone and tooth formation in their young.
- Heavy milking animals need calcium and phosphorus to replace what is going into the milk. (Each litre of cows's milk contains 1.5 g calcium and 2 g phosphorus.)
- Cows turned out to grass in the spring need extra magnesium.
- Young pigs need extra iron in their first 2 weeks of life.
- Local shortages of magnesium and cobalt are sometimes found.

Vitamins for Livestock

Very small amounts of these complicated chemicals are needed by the body. Most farm animals get what they need in the normal rations but, as with minerals, there are cases where trouble is caused by a shortage of one or more vitamins.

Vitamin A helps the animal to keep healthy and to resist disease. It is formed in the animal body from the yellow substance carotene which is found in green foods, carrots, maize and milk (particularly in colostrum).

Vitamin B is a group of several different substances—thiamin (b_1), riboflavin (B_2) and others. It is needed for the nervous system, and is found in the whole grains, wheat feed, grass, milk and yeast. Cattle can make some of their own supplies of this vitamin in the rumen (see Figure 24), but it is needed in the rations of pigs and poultry.

Vitamin D is important for young animals as it is concerned with the proper use of minerals and bone formation. Sunlight on the skin of animals forms this vitamin, and it is also found in hay.

Vitamin E is concerned with breeding. It is commonly found in green food and cereals. Deficiency may cause muscular dystrophy in calves and lambs.

When Extra Vitamins are Needed

There are two main reasons for a shortage of vitamins:

1. Animals reared under artificial conditions and not getting grass or other green foods may suffer from a shortage of vitamin A. This particularly applies to young stock.
2. Young animals under winter conditions or being reared in buildings where they are not exposed to sunlight may suffer from a shortage of Vitamin D.

How to Feed Vitamins

It is a common practice to give vitamins A and D together to young animals. This is cheap and easily done. These vitamins can be supplied in vitamin supplements added to the normal rations.

FOOD QUALITY

There are a number of points about foods which are commonly mentioned. It is important to understand them.

Palatability means whether or not an animal likes the food. If a food is palatable, animals will eat it without trouble. This covers taste, texture (rough, smooth or hairy), dryness, wetness or dustiness and variety—all of these are important. Adding flavourings, such as molasses, to foods makes them more palatable.

Digestibility is important, particularly with young stock or with animals which must get the most from their food without too much work for their digestive system, such as a high-yielding dairy cow. If a food is digestible, it is easily broken down and its goodness extracted; if it is indigestible, it takes much more effort for the animal to do this, and much of it may be wasted.

Bulk refers to the amount of space that the food takes up. The animals with simple digestive systems—pigs and poultry—cannot deal with much bulk. The ruminants have plenty of room for bulk, but it becomes a problem with a heavy yielding dairy cow which has to eat a great deal of food in order to cover all her needs; in this case, bulk must be restricted.

Cost and value The true value of a food is not always the same as its cost. It depends on the feeding value of the food compared with other foods which are available at the time. Thus grains can be compared according to their energy value, and oil-cakes according to their protein value.

FARM FOODS

The Main Groups of Foods

There are many different foods used on any farm. Even if most of the food used is bought ready-mixed—for example, compound foods produced in cube or nut form by one of the feedingstuff firms—it is made up of several different foods. These foods vary in many ways— size, shape, weight, taste, food value, etc. It is only by grouping them, and having some knowledge of what they contain, that we can begin to understand their proper use.

Roots (sometimes called succulent foods) contain a lot of water, some energy substances (sugars and starches), and very little of anything else. They are palatable and filling; usually with a laxative effect. The feeding value is low in relation to their bulk.

Green foods (from the forage crops) have plenty of water, some energy value, but usually more protein than roots. When young they are high in protein; when old the amount of fibre in them increases. These are palatable, filling foods and fairly cheap.

Fodder means hay and straw, made from green crops which are harvested when fully grown (straw) or cut earlier and dried (hay). These foods contain very little water, much fibre and some energy. There is very little protein in straw but varying amounts in hay, as shown on page 68. Fodder is filling, and has a binding effect on the digestion.

Concentrates contain plenty of food value in a small bulk, because they contain very little water, and most of them have only small amounts of fibre. Some are high in energy (the grains), some high in protein (the oil-cakes), and some have reasonable amounts of both. The palatability varies. These are the most expensive foods and must be used carefully.

Additives are any other substances which are put in foodstuffs for special purposes. They include minerals, vitamins, spices and flavourings, and medicines (including antibiotics).

Minerals are essential in small amounts for various body processes, particularly for bone formation and milk production.

Vitamins are essential in very small amounts for the chemistry of the body.

Analysing Foods

It is possible to analyse any food in the laboratory to give an idea of its feeding value. This information can be used in rationing. Figures are quoted in various reference books and can be found on the sack and on the delivery note or ticket of most common foods. These figures include:

- *Dry matter* (DM)   The amount of solid material in the food, apart from the water it contains. Dry matter is high in concentrates, low in water foods such as silage or roots.
- *Metabolisable energy* (ME)   A measure of the energy part of the food
- *Digestible crude protein* (DCP)   A measure of the protein part of the food.

To check on analysis of the less common foods (such as some types of oil-cakes) and feeding value look them up in one of the reference books and compare them with other foods of the same type.

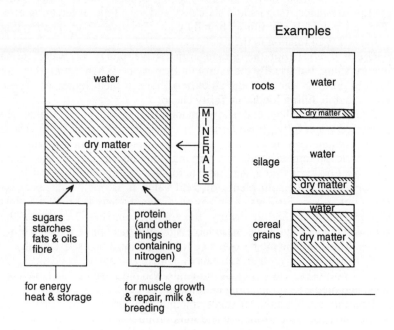

Figure 21   The make-up of food

Roots

Turnips have a very low feeding value, and are commonly fed in the field to sheep. If cows have too much, it may cause a taint (flavour) in the milk.

Swedes have a slightly higher feeding value and keep better than turnips, so are often lifted and clamped.

Mangels are high-yielding and palatable roots but must be ripened before they are fed; otherwise they cause upsets. In practice, do not feed before Christmas, and feed very carefully to in-lamb ewes.

Fodder beet are of various types. The larger are just like mangels and the smaller, which grow deeper in the ground like sugar beet, have a higher feeding value.

Potatoes have the highest feeding value of any of the roots. When fed whole to cattle there is some danger of choking, so they may be chopped or pulped. For pigs, it is best to boil them to make them more digestible. Green potatoes are poisonous if fed raw; cooking makes them safe.

Green Foods

These have more protein than roots. Most of them can be preserved as silage for feeding later, and some can be made into hay.

Kale is the commonest of the green crops (other than grass) and is widely used for autumn and winter feeding, particularly for dairy cattle. The highest feeding value—most of the protein—is in the leaf rather than the stem. *Marrow-stem kale* produces most weight per acre; *thousand-head-kale* is leafier and hardier; it is used later, often into the near year.

Cabbage is very similar in feeding value to kale; it cannot be stored for any length of time.

Rape is nearly all young leaf and so is high in protein. It is good food for sheep, but often they do not do well after going off rape on to any other food.

Sugar-beet tops are a useful food, with the same feeding value as kale. It is important not to feed them fresh, as they contain a chemical which can cause stock to scour. Either let them wilt for a week or more, or sprinkle ground chalk on them if they must be fed fresh.

Grass is a basic, cheap food, whose quality varies from very poor to excellent. Its feeding value depends on a number of things:

- what grasses, clovers and other plants are present in the sward
- whether the grassland is well limed and fertilised
- the age of the grass. The younger and shorter, the more protein there is and the more digestible it is. The older and longer the grass, the less protein and more fibre it contains, and it becomes less digestible.

Silage can be made from grass and several other green foods. It is a sort of pickling process, using acids produced by bacteria working on the green crop. Silage cannot be of better value than the material used, and some of the feeding value is lost in making it into silage. Like grass itself, the quality of silage varies. Also, it depends on how well or badly the silage is made. Tests of silage quality are:

- *Dry matter* (DM) A practical method of testing is:

	per cent dry matter
Water easily pressed out by hand	less than 18 per cent
Water only just pressed out	19 to 24 per cent
Water cannot be pressed out	25 per cent or more

- *Colour* If dark brown, it is overheated; it tastes good but much of the feeding value is lost. If medium green, it is ideal. If light green, it is underheated and may not be palatable to stock.
- *Smell* Rich and tobacco-like, it is overheated. Fruity and acid like pickled onions, it is just right. Very strong, like dung, it is underheated.

Fodder

This is dry food, with a lot of fibre.

Hay is very variable in quality. Its feeding value depends on the value of the grass it is made from (see above). The later it is cut, the bigger the yield but the lower the feeding value and digestibility. How well and how quickly the grass is made into hay also governs the quality. To test the feeding value of hay, look at these points:

- *Colour* Good hay is green; poor hay looks bleached
- *Leafiness* The more leaf, the higher the feeding value; if there is a lot of stem, it is lower in value.
- *Smell* Good hay should have 'a good nose'—a nice fresh smell. Musty hay is of poor quality and can cause trouble.

Straw is commonly fed to cattle, calves, and sheep. It has a much lower feeding value than hay, with little protein, but is a useful feed when clean and palatable. Barley straw is normally used, with some oat straw. Wheat and rye straw have no feeding value. There are chemical treatments of straw, and additives such as ammonia, to improve its feeding value.

Concentrated Foods

These contain little water (high dry matter) and plenty of feeding value. The individual foods are called straights, which can either be fed on their own or mixed with other foods.

Compounds are the ready-mixed foods which are sold by merchants and which are commonly used on farms today. They are made by a number of compounding firms—national and local—and are designed for different classes of livestock and for various conditions. They are ready for use—balanced for their particular purpose—and sold under such names as dairy nuts, sow-and-weaner meal, and ewe and lamb mixture.

Concentrates are also supplied by merchants, but they are not fed direct to livestock. They are designed to be used by the farmer for mixing with his own feedingstuffs, such as home-grown cereals. These foods are sold under such names as dairy concentrate, beef concentrate, and pigbreed concentrate.

Straight concentrated foods fall into four main groups:

1. High energy foods These are mainly the cereals. Barley is the chief cereal fed to livestock in Great Britain. It is a good food for all classes of stock and is mostly used for pig feeding. It should be finely ground. Maize is a little higher in energy value, and is used in the same way. As it contains oil, it should not be used too much in pig rations, where it causes soft body fat. Flaked maize (like corn flakes) is cooked and rolled; it is more digestible than ordinary maize and most animals find it very palatable. Milo is imported and has an energy value slightly less than maize; it is not very palatable. Wheat contains more protein than the other cereals. It should be coarsely ground; otherwise it goes very pasty in the stomach, and it should be used with care. Rye is used in the same way, but not much liked by most stock. Oats contain more husk than the other cereals and is usually fed crushed. Commonly fed to cattle and sheep, it contains too much fibre for pigs. Sugar-beet pulp has just the same feeding value as oats when fed dry; it can also be fed after soaking as a substitute for roots. Other foods now available include 'nutritionally improved straw' and 'cereal pellets' made from imported starchy foods.

2. Milling by-products These foods have more protein and fibre than the cereals. They were previously common foods on the farm but are less used today.

Wheatfeed (also known as middlings or weatings) is a by-product of wheat milling—what is left after the white flour is produced. It is used in cattle and pig feeding, and helps to keep the digestive system in condition. Bran is used more as a medicine than a food. Fed as a bran mash—particularly to pregnant animals—it prevents constipation.

3. Medium protein foods These contain reasonable amounts of both energy and protein. Peas and beans are used less today for livestock feeding. When ground they are a useful food. Dried grass and dried lucerne are used in many mixtures of feeding stuffs; they supply part of the carotene which animals need to form vitamin A in the body; it also contains a fair amount of fibre. The protein content varies according to the stage of the grass when cut.

4. Protein foods   There are three main types:

- *Animal proteins* have a higher feeding value, but are expensive and sometimes in short supply. They must be used with care particularly if they contain offal.

    White fish meal is made from whole 'white fish' (non-oily) which are ground up, bones and all. It has a high protein value and also supplies useful minerals. Herring meal is oilier, higher in protein, and has a stronger flavour. Meat-and-bone meal is lower in protein content than either type of fish meal, and there is much less mineral matter in it. Milk powder (or dried milk) is an expensive product, only used for young animals—as in rearing calves or lambs.

- *Vegetable protein foods* are mostly oil-cakes and meals made from oil seeds after the oil has been removed by crushing or extraction. There are a number of these foods, but some are only used by food compounders in their products, and there are very few of these foods now available to farmers.

    Soya bean meal, which is one of the highest of these foods in protein, is commonly used for mixing. Linseed cake is expensive, and is sometimes used for special purposes; it has a laxative effect and makes cattle look well by putting a 'bloom' on their skin. Brewers' grains are a by-product of brewing. They can be bought by farmers from local breweries and are sometimes used for dairy cows—either as 'wet grains' or as 'dry grains'. This food is balanced for milk production. Wet grains do not keep for long and must be used up quickly or can be ensiled and stored. Some yeast products are available as a by-product of brewing, and can be used in pig rations.

- *Artificial nitrogen foods* are now in use, mainly provided by merchants as part of a compound food, or incorporated in supplement blocks or liquids which are fed to grazing animals to balance their bulky foods. Urea is the best known of these.

Comparing Foods

It is sometimes useful to be able to compare foods, to know which are the best to provide certain food constituents. The table on page 71 shows how common farm foods compare, with the highest quality first and the lowest last.

Dry matter	per cent DM	Energy	ME	Protein	DCP
Oil-cakes and concentrates		Maize	___	Fish meal	
Cereals	___	Linseed cake		Meat meal	___
Straw	___	Other oil-cakes	___	Groundnut cake	___
Hay	___	Barley	___	Soya-bean meal	___
Silage	___	Wheat	___	Dried yeast	___
Grass	___	Peas	___	Cottonseed cake	___
Kale and rape	___	Beans	___	Linseed cake	___
Mangels	___	Oats	___	Beans	___
Swedes		Sugar-beet pulp	___	Peas	
Turnips	___	Groundnut cake	___	Coconut cake	
Cabbage	___	Wheatfeed	___	Palm kernel cake	___
		Hay	___	Dried grains	___
		Potatoes	___	Wheatfeed	___
		Silage	___	Grass	___
				Silage	___
				Hay	___

(Complete these tables, getting the figures from a reference book.)
(See top of page 66 for the meaning of DM, ME and DCP.)

DIGESTIVE SYSTEMS

Food is the most important single thing we are concerned with in keeping farm livestock. And the part of the body which deals with the food—the digestive system—is the largest and in many ways the most important part of the animal body when it is alive.

Basically, this digestive system is a sort of machine to get food into the body, to break it down, and to take the goodness from it so that the body, with all its other systems, can use it for various purposes.

All our main groups of farm animals have different digestive systems—pigs, poultry, horses and the ruminants (cattle and sheep)—but each digestive system has something in common with the rest, and we can get the general idea by looking first at simple digestive systems.

Simple Digestive Systems

This is the system found in the pig, which is very similar to that in our own human bodies. It is suitable for dealing with a mixed diet—vegetable and animal foods—without too much fibre or too much bulk.

The mouth   Food is taken into the mouth, pulled off or bitten into lumps by the front biting teeth (incisors) and chewed up by the back grinding teeth (molars). During this chewing it is mixed with the liquid saliva which comes from various glands beneath the tongue and elsewhere in the mouth: saliva contains a chemical (an enzyme—ptyalin) which starts to break down the starches and sugars in the food.

The food passes back over a sort of trapdoor (the epiglottis) which prevents it going down into the lungs (which would cause a good deal of trouble). It goes into the gullet (oesophagus), which is a long tube leading into the stomach.

The stomach is a large container, which varies in size according to the amount of food in it; it is not just a simple bladder like a balloon, but a very sensitive living organ which churns its contents about and has several jobs to do. In the stomach, the food is mixed with digestive juice from the

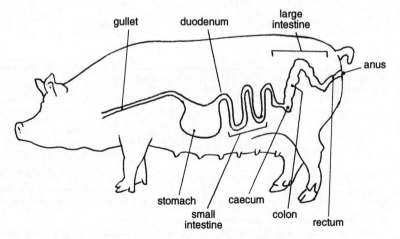

Figure 22   Simple digestive system in the pig

glands in the stomach wall. This acid juice starts to break down the protein part of the food. When the stomach has worked on the food long enough, it passes it on, through a muscular ring (a sphincter), to the next stage.

The small intestine is a very long tube. In the first part of it (the duodenum) more digestive juice is mixed with the food. This juice comes partly from the liver (bile) and the pancreas (pancreatic juice). The food passes along into the two next portions of the small intestine, which is very looped and curled around inside the body. More digestive juice is produced from small glands in the wall of the intestine, and as food passes along, it is broken down chemically into a much more simple form. Also, as it passes along, some of the food material is taken out and passed into the bloodstream.

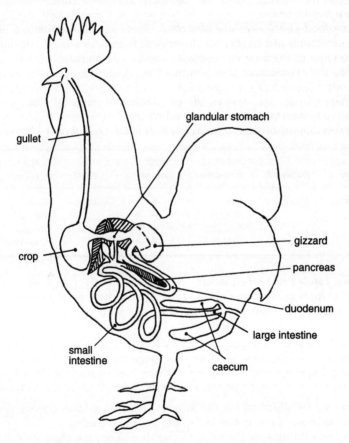

Figure 23 Simple digestive system in the chicken

The large intestine is a shorter but much broader tube. At the beginning of it is a small extra piece, a sort of blind alley leading to nothing. This is the caecum, and in it bacteria break down some of the fibre in the food. In the large intestine the broken-down food substances are taken out and pass into the bloodstream, and water is removed. The food is thus getting drier and more solid, until at the end only the indigestible parts are left, which cannot be broken down any more or used by the body. This rubbish, mixed with some of the waste products of the body, passes out of the body as dung, at the anus (another sphincter, which opens when necessary).

How food moves along   The gullet, the stomach and all parts of the intestine are muscular and move. By contracting and expanding they push the food along, rather like the movement of a snake—an action known as peristalsis.

How food substances are absorbed   Parts of the intestine are lined with tiny things like fingers which stick out from the walls into the liquid and broken-down food as it passes. These villi contain blood vessels, and the food substances pass into them and so into the blood system of the body.

Where foods are broken down   Different parts of the simple digestive system work on different substances in the food, so that in the end everything is broken down as much as it can be. It works like this:

Mouth	Stomach	Duodenum	Small intestine	Caecum	Large intestine
Starches	Proteins	Fats ➤ ➤ ➤ ➤ ➤		Fibre	Water
		Sugars ➤ ➤ ➤ ➤			and
		Starches ➤ ➤ ➤ ➤			Foods
		Proteins ➤ ➤ ➤ ➤			Absorbed

How foods are broken down   Most foods start off as a mixture of rather complicated substances. The digestive system breaks them down into the simplest form, which the body can then use. It happens like this:

Sugars➤➤➤➤➤➤➤➤➤➤➤➤➤➤➤➤Simple Sugars➤➤➤➤➤➤Glucose
Starches➤➤➤➤Sugars➤➤➤➤➤➤➤➤Simple Sugars➤➤➤➤➤➤Glucose
Fats and Oils➤➤➤➤➤➤➤➤➤➤➤➤➤➤➤➤➤➤Fatty Acids and Glycerol
Proteins➤➤➤➤➤➤➤➤Simple Proteins➤➤➤➤➤➤➤➤Amino Acids

This leaves fibre, which the simple digestive system cannot really tackle, so most of it is passed out of the body in the dung.

Enzymes do most of the work of digestion. They are chemicals in the digestive juices which break down food substances and help to change one substance into another. One simple example is the enzyme in saliva, which changes starch into sugar.

The chicken's digestive system is very similar to that of the pig, and it can deal with the same types of food. The stomach arrangements, however, are rather different. The food is taken down the gullet into the crop, where it stays for a while as it is soaked and softened. It passes on through the glandular stomach, where it mixes with digestive juices; it is released from there in small amounts. Next it travels into the gizzard, which is a strong muscular container, with horny walls. The gizzard contains small pieces of grit and stones; the chicken has to eat hard grit once in a while to top up its supply of these. In the gizzard, food like grain is ground up and mixed thoroughly with the digestive juices. It then passes on to the duodenum and the intestines.

Complex Digestive Systems

Pigs, poultry and humans eat a mixed diet, but they cannot really deal with much of anything very bulky (such as roots or grass) nor anything very fibrous (hay or straw). The more complicated digestive systems are found in those animals which naturally live on such foods.

Rabbits have a curious system; everything goes through them twice in order to take all the goodness from the food. The rather soft dung produced after the first time through is eaten, and then goes through the animal again. Finally, all that is left is the little balls of very dry dung which we find on the ground or in the pen—these consist of very hard or indigestible material.

Horses have a small stomach which can be upset rather easily by wrong feeding methods. The fibre in their diet is digested mostly in the very large caecum, where it is attacked and broken down by bacteria. This system produces plenty of gas, which passes out of the animal at the rear, and the food substances are absorbed into the bloodstream in the large intestine, which also is very big.

Ruminants are a group of animals which include cattle, sheep, goats and deer. They all have a complicated stomach, specially designed to deal with fibre.

The ruminant stomach is in four compartments. The first and largest is the rumen (or paunch). It is always partly filled with water, and the food which comes down the gullet goes into it and soaks. While there it is worked on by bacteria, and broken down. When the animal is chewing the cud (ruminating) it pushes lumps of food back up the gullet into its mouth, and chews them very thoroughly. This food then returns into the rumen. A cow may spend up to 8 hours out of very 24 chewing the cud.

All the food in the rumen is soaking away in a sort of thick porridgy mass, being churned around by the muscular walls. Conditions are wet

HORSE

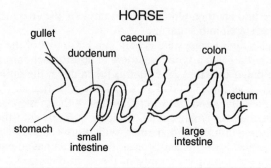

RUMINANT

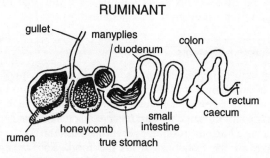

Figure 24   Complex digestive systems in the ruminant and horse

and warm, ideal for bacteria, several different types of which work on the food in the rumen. As well as breaking down the food substances, they produce gas and this is belched out by the animal every now and then. Various things can cause this system to stop working properly; then the gas cannot get away and this causes bloat—the rumen gets full of gas, which causes pain and can lead to serious trouble in a bad case.

The rumen is connected directly to the second compartment, the honeycomb (reticulum), which acts as a sort of filter and holds water and hard objects which are swallowed. Food which is chewed or ground up fine enough passes through a small hole into the next compartment. Large pieces are returned for more treatment.

The food passes through into the manyplies (omasum), which is partly filled with flat leaves of tissue. This compartment acts as a filter, churns the food more, and takes out much of the water in the food. The food is thus much drier when it gets into the true stomach (abomasum), where the normal digestive juices can work on it.

It should be noted that the young ruminant (such as the calf) has only a simple stomach (the true stomach), and the other compartments of the stomach do not develop until weaning time. This means that calves cannot digest fibrous foods at first, and so need milk and other concentrated foods.

The food then passes on to the duodenum and there, in the small and the large intestines, it is digested in very much the same way as in the simple digestive system.

As ruminants have plenty of storage space in relation to their size, and can eat their food quickly and then chew it very thoroughly in their own time, they are able to deal with the bulky and fibrous types of food very efficiently. But they can also deal with grains and other concentrated foods.

This table shows how the various digestive systems of farm animals compare:

	Capacity of stomach (litres)	Capacity of whole digestive system (litres)	Length of intestines in comparison with body length
Pig	9	27	15 times
Sheep	23	36	26 times
Cow	180	300	20 times
Horse	18	200	15 times

RATIONS

There are several systems of working out rations for farm animals—some of them very complicated, some quite simple. First of all, get some idea of the way the main types of farm animals are fed in practice, before thinking about how to work out rations.

Pigs are mainly fed on ready-mixed compound foods—made of concentrated grains and protein foods. Pig foods are often mixed on the farm. Small amounts of green foods or roots may be fed.

Poultry are fed on ready-mixed compound foods, with some whole grain.

Cattle are fed mainly on bulky foods produced on the farm, with concentrates according to the amount of milk or growth they are producing. *Calves* are fed on milk or milk-substitute at first, later on with concentrates and some bulky foods which are gradually increased in quantity.

Sheep are fed largely on grass, with other bulky foods in the winter, and small amounts of concentrates later in the winter.

Working Out Rations

In the next few pages you will be shown how to work out simple rations for the main classes of livestock. These methods are used in practice, and although they are not completely accurate they can always be checked with reference books and with advisers from the Ministry of Agriculture (ADAS) and from feed firms.

The first step—particularly with cattle, where bulky foods are being fed—is to make sure of the quantities of foods available. From this you can work out a daily or weekly ration. This table shows how to work this out:

Total quantity of food available per head (tonnes)	Daily rations (kg)			
	3 months (90 days)	4 months (120 days)	5 months (150 days)	6 months (180 days)
1	11	8.3	6.6	5.5
2	22	16.6	13.2	11
5	55	41.5	33	27.5
10	110	83	66	55

These simple rationing systems are satisfactory up to a point; rations must be worked out by common sense, with some idea of what the animal needs at its different stages of growth and according to the job it is doing. You can check on these rations by calculating the animal's needs and comparing them with the feeding value of the foods in the ration.

Stockmanship

The best and first check on a ration is to see if the animal is 'doing well' on it. This requires experience and good observation. The animal should be healthy and content, and its dung should be in proper condition—neither too loose nor too firm.

RATIONS FOR PIGS

Food makes up about 80 per cent of the cost of a pig, and feeding is the most important part of the management of fattening pigs. It has a great effect on good growth of any pig, it is particularly important with the young pig (on creep food), and the size and strength of litters is influenced very much by feeding the sow.

The pig has a small stomach and a simple digestive system. It cannot deal with much bulk or fibre. Fattening and growing pigs are putting on weight quickly and need good quality digestible food.

Pigs are fed largely on meal mixture, and their rations are normally given in kg (or pounds) of meal per day per pig.

Meal can be fed wet or dry, ad lib (as much as they want) or restricted. It may be mixed on the farm or bought ready compounded.

Daily Rations

These vary according to the size of pig, time of year, type of management, grazing available, etc. These figures give some idea of average rations per day:

Young breeding stock	1.4–2.7 kg	(3–6 lb)
Pregnant sow	1.8–3.6 kg	(4–8 lb)
Suckling sow	3.6–5.4 kg	(8–12 lb)
Full-grown boar	1.4–3.0 kg	(3–7 lb)
Fattening pigs	0.9–2.7 kg	(2–6 lb)

Type of Rations Needed

Creep feeding   ad lib, high quality, palatable, digestible. Pellets are usually better than meal.

Weaner   ad lib, high quality, plenty of protein.

Pork or bacon pig   a weaner ration, fed ad lib, until at a certain weight (which varies with the type of pig) the ration is restricted (at 45–55 kg liveweight).

Breeding stock   same as the weaner ration, with plenty of protein. Some bulky foods such as grass, roots and green food can be used instead of part of the meal ration.

Meal Mixtures

In practice, many farmers use three mixtures: Creep feed, sow and weaner ration and fattening ration.

Most pigs are fed on bought-in compound foods, which are mixed and sold by many different firms both national and local.

Farmers also mix their own foods, using home-grown cereals (chiefly barley) balanced with various forms of protein-rich food.

Substances containing vitamins and minerals are commonly added to pig foods. In any pig food, both oil and fibre are kept at a low level.

Farm foods commonly used for pigs include:

Cereals—Barley, maize wheat and oats
Cereal by-products—Wheatfeed (middlings) and bran
Vegetable protein—Soya-bean meal, peas, beans, grass and lucerne meal, yeast
Animal protein—Fish, meat meal, whale-meat meal, separated milk

Grazing is valuable, so long as it is of good quality. Some special clovery leys are grown for pigs.

Water is important for pigs, as for all livestock. The amount needed varies from 3.5 litres at 8 weeks old to 7–9 litres at 6 months. A sow in milk may need up to 12 litres.

Example of sow and weaner ration

Cereals	60%
Wheatfeed	30%
Vegetable protein	$2\frac{1}{2}$%
Animal protein	$7\frac{1}{2}$%
plus minerals and vitamins	

Example of fattening ration

Cereals	75%
Wheatfeed	15%
Vegetable protein	$7\frac{1}{2}$%
Animal protein	$2\frac{1}{2}$%
plus minerals and vitamins	

Foods commonly used in mixtures are as follows:

Cereals

	Maximum limit in ration	Grinding
Barley	90%	fine
Maize	30%	medium
Wheat	60%	coarse
Rye	25%	coarse
Oats	20%	fine

Cereal by-products Middlings is commonly used in pig rations. Farmers are used to it, it lightens a mixture and it is good for digestion.

Bran has a special value for sows at farrowing time, when they are liable to be constipated.

Vegetable protein foods The commonest now is soya-bean meal. Substitutes for it are:

	Value*	Maximum limit in the ration
Dried yeast	1	5%
Peas, beans	2	15%
Grass meal, lucerne meal	$2\frac{1}{4}$	25%

  * (as a substitute for soya-bean meal)

Animal protein foods The commonest is white fishmeal. Substitutes for it are:

	Value*	Maximum limit in the ration
Meat meal	1	10%
Whale meat meal	1	10%
Meat and bone meal	$1\frac{1}{4}$	$12\frac{1}{2}$%
Separated (skim) milk		3.5 litres per day

  * (as a substitute for white fishmeal)

Minerals Ready mixed minerals should be fed at makers' recommendations. A simple farm mixture can be made up as follows:

3 parts limestone flour ⎫
1 part salt   ⎬ at $2\frac{1}{2}$% of ration
1 part steamed bone flour ⎭

  (If feeding fishmeal at more than $7\frac{1}{2}$ per cent of ration, leave out the steamed bone flour.)

Vitamins A and D are commonly fed, either as synthetic vitamin supplements or as cod liver oil.

Bulky foods Within reason, the following foods can be used to replace cereals, as follows:

Equal to 1 kg of barley meal: 4 kg cooked potatoes
                                  3–4 kg cooked swill
                                  5–6 kg fodder beet
                                  6–7 kg green food
                                  12 kg whey

These foods are used sometimes for breeding stock, less commonly for fattening pigs.

RATIONS FOR POULTRY

The normal practice in poultry feeding is to provide a compound food ad lib for layers. This contains cereals mixed with high quality protein foods, minerals and vitamins A and D. It is common to feed ready mixed compounds, which are supplied either in meal or pellet form. Poultry should also have plenty of water, and should be able to help themselves to hard grit.

Meal for poultry can be mixed on the farm. A bought-in 'pre-mix' containing all the necessary protein, minerals and vitamins can be added to cereals, or the whole ration can be mixed on the farm. If the ration is farm-mixed, it should consist of cereals, animal and vegetable protein foods, minerals and vitamins, with an analysis of 15–18 per cent protein.

Example

3 parts wheat	1 part fishmeal
3 parts maize	1 part vegetable protein foods
1 part wheatfeed	plus minerals and vitamins

An average hen eats up to 50 kg of food in a year but intake varies from 85 to 140 grams per day. The modern hybrid hen has a small capacity and needs highly concentrated foods.

RATIONS FOR CALVES

Details of rearing methods and specimen rations are given on pages 7–9. The feeding of calves is closely tied up with systems of management, and depends upon the purpose for which they are reared—veal, beef or dairy.

Beef Young stock being reared for beef production need good quality foods of sufficient quality to ensure steady growth without any checks.

For early beef (up to 18 months old) there must be no store period— thus good feeding is needed all the time.

Dairy The dairy heifer need not be fed too well. The expense of high feeding is not justified—she will milk no better later and may get into too good a condition for breeding.

Below is an example of daily feeds for dairy heifers, but there are several different systems.

6 to 12 Months Old

- Up to 0.5 kg (1 lb) good hay per month of age (partly replaced by equivalent foods—see page 84)
- 1–2 kg ($2\frac{1}{2}$–$4\frac{1}{4}$ lb) of concentrates daily
- Suitable compound foods or home mixtures which will vary according to the quality of the bulky foods.

When young stock go out in the spring and summer on to good grass, they will manage without any extra feeding.

12 Months Old Onwards

- 0.25 to 0.35 kg ($\frac{1}{2}$–$\frac{3}{4}$ lb) good hay per month of age (partly replaced by equivalent foods if necessary).
- 1 kg ($2\frac{1}{4}$ lb) or more of concentrates daily.

On good grass, no other feeding will be necessary.

RATIONS FOR DAIRY COWS

It is common to consider the ration in two parts:

Maintenance ration The food needed to keep the cow in good health and condition. Balance of protein to energy is about 1 to 10. The amount needed varies with the liveweight of the cow.

Production ration The food needed to provide for the production of the cow (milk, a calf, preparation for milking). Balance of protein to energy is about 1 to 5. The amount needed varies with the actual or expected production of the cow as measured in litres per day.

Summer Feeding

Grass will provide for the normal maintenance requirements of dairy cows. It may also cover the production of a certain amount of milk, depending on the quality of the grass, the way it is managed and the time of the year. You should aim for as much production as possible from grass, as it is cheaper for cows than most other foods.

Young, high quality grass contains enough protein and not enough energy for the cow's needs and it can be balanced by feeding some cereals; a little hay should also be fed to provide some fibre.

Winter Feeding

By feeding plenty of bulky foods it may be possible to provide maintenance and a good part of the production ration also. This needs good quality foods and the right sort of cows. It is sometimes done by self-feeding silage. A more common method is to provide maintenance with bulky foods, and to feed concentrates for production.

A simple system for working out a maintenance ration is the use of hay equivalents. A cow needs so many kg of medium quality hay per day, according to its size.

	Liveweight (kg)	Hay per day (kg)
Friesian	560	9
Ayrshire	460	7.5
Jersey	380	6.8

One kg medium quality hay can be replaced by any one of the following:

0.6 kg oats or beet pulp
0.75 kg dried grass or good hay
3 kg silage
3 kg fodder beet or potatoes
4 kg kale, beet tops or swedes
5 kg mangels or wet beet pulp
0.5 kg straw and 3 kg kale, beet tops or swedes

Production ration is usually based on the feeding of 0.4 kg of balanced compound food to produce 1 litre of milk of average quality. For milk of higher quality (such as Channel Island) approximately 0.5 kg will be needed.

Examples of balanced milk production rations

Dairy cake or good dried grass
$\frac{1}{2}$ beans and $\frac{1}{2}$ oats (or beet pulp)
$\frac{1}{2}$ grain balancer cake and $\frac{1}{2}$ cereals
$\frac{1}{4}$ high protein cake and $\frac{3}{4}$ cereals

For special purposes, there are more highly concentrated foods which are fed at a lower rate

Steaming-up means special feeding for the cow before it calves, to prepare for the next lactation. We start to feed a balanced dairy ration 6 weeks before calving and gradually increase up to a daily ration just before calving which would cover $\frac{3}{4}$ of the expected yield. For high yielding cows this method needs very careful management, and may cause the cow to milk before calving.

Water   A dairy cow needs about 22 litres per day. In addition a cow in milk may need up to 3 litres of water for each litre of milk being produced.

Minerals   There are about 6 g of mineral matter in each litre of milk. Ample minerals must be provided in the ration (add to home-mixed foods).

Dry matter   A cow can deal with 1.4 kg (3 lb) of dry matter for each 50 kg (1 cwt) of her body weight. Keeping the amount of dry matter down is sometimes a problem with high yielding cows.

Some well-known breeds of cattle

FRIESIAN/HOLSTEIN
A machine for producing milk (and much of our meat). The black and white cow which has conquered Europe.

Plates by Simon Tupper except where indicated otherwise

JERSEY
A quality cow which produces milk with high butterfat. Small, efficient and productive. A main source of milk products.

HEREFORD
Meat from grass. One of the great beef breeds—either pure or for crossing with other breeds.

ABERDEEN ANGUS
A quality beef breed. Hardier and slightly slower in growth than some other beef breeds. A cross for quality meat and for easier calving.

CHAROLAIS
An imported breed which has become established quickly. Used for crossing, to produce quality, lean meat from the dairy herd.

Some well-known breeds of pig

LARGE WHITE
Long and lean—the real bacon pig.

BRITISH SADDLEBACK
A deep-bodied, good mothering type of sow, with big litters and high milk yield.

DUROC
A recent hybrid cross pig, bred from European stock, for body shape improvement and good growth rate.

Some well-known breeds of sheep

SUFFOLK
One of the leading meat breeds, commonly used to produce finished lambs. Meaty and fast growing.

TEXEL

An important meat breed, fairly recently imported. Blocky, meaty, fast growing and lean.

NORTH OF ENGLAND MULE SHEEP
An example of the commonest breeding sheep, used for crossing with a meat type ram to produce finished lambs.

SCOTTISH BLACKFACE
A breed which makes best use of poor land. The start of the breeding chain in which, after several generations, crossing leads to a finished lamb (see 'stratification', p. 30).

Fibre Every cow in milk needs a certain amount of long fibrous food such as hay. Even the highest yielders, which do not have much room for hay because of the concentrates they need, should not have less than 3 kg (7 lb) of hay each day.

Digestion The general effect of the ration should be slightly laxative. Palatability of foods is sometimes a problem.

Farm foods commonly used for dairy cows:

Fodders	Hay (meadow, seeds, lucerne), straw (oat, barley)
Green foods	Grass, clover, lucerne, kale, cabbage, rape, maize, sugar beet tops, silage
Succulents	Mangels, fodder beet, swedes, turnips, potatoes, wet beet pulp, wet brewers' grains
Concentrates	Dairy cake, oil-cakes and meals, fish meal, beans, peas, dried brewers' grains, beet pulp, cereals, dried grass

RATIONS FOR BEEF CATTLE

Maintenance As with the dairy cows, rations for some beef cattle can be worked out simply by the hay equivalent method (see pages 83–84). A store beef animal needs 1.5 kg of medium quality hay for each 50 kg liveweight, to cover maintenance and steady growth (without fattening). Examples:

Liveweight (kg)	Hay per day (kg)
300	9
350	10.5
400	12

Use any of the ordinary bulky farm foods to substitute for part of the hay needed, and to give a varied ration that the animal will like. Beef cattle do not eat with such a good appetite as dairy cows, and sometimes have to be tempted with tasty food.

Grazing During the growing season, grass will provide all or most of the needs for maintenance and steady growth, according to its quality. Good grass will also fatten beef cattle.

Production For finishing, concentrates are needed as a rule— although it is possible to fatten cattle on bulky foods, if you include better quality foods such as hay and silage, instead of just straw and roots.

About 1.8 kg (4 lb) of concentrates (or other dry food) will produce 0.5 kg (1 lb) of liveweight gain. So for a good finishing rate of 1 kg (2 lb) liveweight gain per day, 3.6 kg (8 lb) of concentrates will be needed.

Young beef More beef is being produced today from young cattle which are finished intensively while still growing. The faster the animal has to grow, the more concentrated must be the ration.

A typical ration under this system would be:

- 17 parts cereals, 3 parts protein and mineral supplement, fed ad lib (plus a small amount of straw)
  or
- maize silage in quantity plus some concentrates

RATIONS FOR SHEEP

Autumn/Winter

Ram Feed well before and during mating season with high protein concentrates.

Ewes Flush 2–3 weeks before they are put to the ram by giving fresh, short, protein-rich pasture. Slowly improve the diet until lambing time, increasing it more during the last 6 weeks.

Flushing is described on page 34.

Grazing may be enough until the end of the year, if the weather is not too hard.

From Christmas (or earlier in bad conditions) daily: $\frac{1}{4}$ to $\frac{1}{2}$ kg ($\frac{1}{2}$ to 1 lb) hay; up to 4 kg (8 lb) kale or silage; $\frac{1}{2}$ kg (1 lb) concentrates (such as 3 parts cereals, 1 part soya-bean meal).

Until lambing, increase rations for the ewe up to 0.7 kg ($1\frac{1}{2}$ lb) concentrates daily. Take care when feeding roots (particularly mangels) before lambing. After lambing, cut out concentrates until lambs are able to take all the milk the ewe is producing.

Example of daily ration from 1 week after lambing until there is plenty of grass:

0.5 kg (1 lb) hay, up to 3 kg (7 lb) roots, 0.5 kg (1 lb) concentrates

Finishing sheep A 35 kg (75 lb) lamb in the autumn needs daily:

0.5 kg (1 lb) hay, 7 kg (15 lb) roots, 0.25 kg ($\frac{1}{2}$ lb) concentrates

Summer

Ewes Feed grass alone. For grazing purposes, 5 sheep = 1 cow. Ewes after weaning are put on poorer pasture.

In feeding the ewe, remember: FAST her (after weaning), FLUSH her (for mating), FEED her (up until lambing).

Minerals should be available for sheep at all times. A good mixture should contain trace elements which may be lacking locally. Minerals may be added to any concentrated feed, fed separately for the sheep to help themselves, or provided as blocks.

Water Sheep need water, particularly when milking, and it should always be available.

Lambs Feed milk from birth, increasing amounts of grass from 2–3 weeks old. Creep feeding of lambs is sometimes done. There are 2 methods:

- *Grass creep* Let the lambs go ahead of the ewes on to fresh short pasture, using a creep-gate in the fence.
- *Trough food* Only the lambs can get at it. Start with 100 g ($\frac{1}{4}$ lb) per day of lamb nuts (or a mixture such as 2 parts oats to 1 part beans) and gradually increase. This food will be needed from 2–4 weeks old until there is plenty of grass.

BALANCING RATIONS

A ration is 'balanced' if the totals of the energy and protein parts of the food it contains are in the right proportions for the needs of the animals which are being fed. This applies particularly to the concentrated part of the ration, where we need to make sure foods are balanced for milk production or other needs.

Details of pig rations are shown on pages 78 to 81. Cattle and sheep are fed on quantities of bulky and fibrous goods (such as hay, straw, silage and roots, and smaller quantities of concentrated foods. It is important that the concentrated foods are balanced for the needs of the animal and in relation to the quality of the bulky foods in use.

In practice, farmers buy in ready mixed or processed foods, or have mixtures made up to their own formula (either on the farm at feed mills). Examples of such foods include dairy cubes, sow and weaner meal, ewe and lamb nuts and various poultry foods.

The foods available for mixing vary from time to time, getting cheaper or dearer according to international market conditions. There are a number of different cereals and other energy foods; by-products such as oil cakes and meals; waste products and other foods not suitable for human use. It is a useful exercise to check occasionally on the foods available and their prices; this can be done from the farming papers and local feed firms.

Many farms grow cereal crops which can be used to make up the main part of a concentrate ration. In the past, it was common to feed a simple mixture such as half oats and half beans as a production ration for dairy cows. Now more complex rations are often fed, consisting of different cereals and energy foods, and other foods higher in protein.

FOOD CONVERSION

The food eaten by farm animals is used partly for maintenance and partly for production. The meat-producing animals—pork and bacon pigs, veal calves, beef cattle, fat lambs and table poultry—use part of their food to produce meat and fat. This is measured by their liveweight gain (their increase in body weight).

We can look on these animals as 'machines to produce meat'. Food is fed into them and they turn it into meat, wasting some of it in the process. Some are much more efficient at this job than others. This efficiency is measured and expressed as a 'food conversion ratio' (see also page 52).

Examples   A pig increases in weight from 20 kg to 45 kg liveweight and takes 75 kg of food to do so. It has gained 25 kg of liveweight for 75 kg of food, which is 3 kg of food for each 1 kg of liveweight gain. We say that its *conversion ratio* is 3 to 1.

The higher the conversion ratio, the more food is needed to produce each kg (pound) of liveweight gain. A higher ratio means the animal is less efficient at the job of converting food into growth. Young animals have a good, low conversion ratio; in older animals it is much higher and thus poorer.

It is easy to measure conversion ratios with pigs as they are usually fed on meal alone. The same applies to poultry, both for eggs and for meat. Conversion ratios are very important, as they have such a great effect on profit or loss.

With cattle and sheep, both of them ruminants (see page 75), the diet often includes foods which are wet and bulky, so it is not at all easy to work out conversion ratios: it is necessary to calculate the amount of 'dry food' eaten by the animal, the dry matter content of all the foods being taken into account.

THINGS TO DO

1. Get to know the whole range of common farm foods available at the present time.
2. Practise estimating quantities and weights of foods and liquids, and compare these with the measured amounts.
3. Calculate daily winter rations for livestock from the quantities of bulky foods available on the farm (see table on page 78).
4. Practise judging the quality and feeding value of hay, silage and grassland.

5. Watch the mixing of concentrate foods on the farm or else visit a mill or factory where compound feed or foodstuffs are made.
6. Find out the analysis and cost of compound foods sold by local merchants, such as sow-and-weaner meal, creep feed, dairy nuts and calf foods.

QUESTIONS

1. What are the main constituents of farm foods? How are the foods grouped together for practical purposes?
2. What particular needs do farm animals have for minerals and vitamins?
3. How do you assess the quality and feeding value of hay and silage on the farm?
4. What are the main differences between the ruminant and the simple digestive systems? What are the main parts of these systems?
5. What is the food conversion ratio? With what sort of livestock can it be worked out most easily?
6. How do you work out a balanced ration for a dairy cow, for a sow and litter, or for a ewe just before lambing?

Chapter 5

Breeding Livestock

MOST OF the production of farm animals depends on their breeding. It is necessary to know and understand the animal's reproductive system and how it works; also to know something about the systems of breeding used by animal breeders and commercial farmers in Great Britain as well as a little about the science of breeding—*Genetics*.

REPRODUCTION

All types of farm animals need to breed—to carry on their kind. This is done by sexual reproduction; two sexes are needed, the male to fertilise the female. Thus the offspring, however many of them there may be, get half their character from the male (the sire) and half from the female (the mother or dam).

Sexual Development

The sexual organs are present in the young animal when it is born, but they are not fully formed. As the animal grows, these organs develop; at a certain age (puberty) there are some body changes and the sexual organs become ready for breeding use. Puberty comes before the animal is fully grown, and an animal should not be used for breeding too early. For this reason, a young boar or bull should not be used too much at first, nor a gilt or heifer mated until it is properly grown. Ewe lambs (in their first year) can be mated, but they are still growing even while producing a lamb.

Sexual Organs

Male There are two glands—the testes—held outside the body in a bag of skin called the scrotum. The testes produce sperms and, along with other glands nearby, produce the fluid—semen—which contains and preserves the sperms. Each sperm is very small, about 0.05 mm long, and several million are produced at one mating. (In birds, including poultry, the testes are inside the body.)

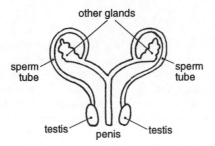

Figure 25 Male reproductive system

Castration is an operation done on male animals which are not needed for breeding purposes. The testes are removed or treated so that they do not develop in the young animal. This makes the animal unable to breed, makes it quieter and less troublesome, and may improve the quality of the carcase.

As it gets older, an uncastrated male animal develops strong flavoured and coarse meat, its body shape alters and often it becomes more aggressive.

Female There are two glands—the ovaries—inside the body. The ovaries produce the female's eggs (ova), usually a few at a time. The ova are small—about 0.2 mm. They are produced from the ovaries at certain times and break away from the surface of each ovary; they pass into something like a funnel nearby and then down the egg tube (fallopian tube) towards the womb (uterus).

The womb is the place where the fertilised egg develops into the young creature (embryo) which later becomes a calf, lamb or piglet, etc. At the mouth of the womb is a ring of muscle known as the cervix. The vagina reaches the outside of the animal at the vulva, which is the lower opening beneath the tail of the female farm animal.

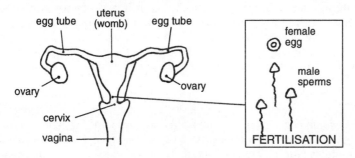

Figure 26 Female reproductive system

Heat Periods

After a female reaches puberty, she starts to have heat periods; this is known as oestrus, the time when the eggs are released from the ovaries. This may be at certain seasons of the year only—as with the ewe and the mare—or right through the year—like the cow and the sow. The length of oestrus varies with the animal; in the winter it is not so noticeable, and is then known as a 'silent heat'. During oestrus, one of the ovaries produces eggs, and hormones released in the blood make the animal feel like mating and make the womb ready for a fertilised egg.

When this happens, the animal is known as 'on heat' or 'in season'. These are common signs of oestrus:

- The animal is restless and excited; a cow or hiefer may bellow.
- She may mount other animals or be mounted by them (bulling or hogging).
- The vulva becomes red and swollen, and there may be a discharge.
- Sows will let the pigman sit astride their backs.
- The animal will allow mating.
- Cow's milk yield may drop and her temperature rise.

| | Heat periods | |
	How long	How often
Sow	2 to 4 days	20 to 21 days
Cow	1 day	20 to 22 days
Ewe	1 to 2 days	16 to 17 days
Mare	5 to 7 days	2 to 3 weeks

Mating

The female will allow mating (usually called 'service' in farm animals) only during the period she is on heat: the male is ready for it at almost any time. Veins in the male penis become full of blood and it stiffens, so that it can be put into the vulva and thus into the vagina. After a time (quickly in cattle and sheep; more slowly with pigs) and some movement, the semen is released and gets through the cervix into the womb.

Fertilisation

In the womb, or in the egg tubes, sperms meet the eggs; one breaks through and buries its head right in the egg. The egg is fertilised when the sperm unites with it. The nucleus of the sperm joins with the nucleus of the egg. The fertilised egg goes to the wall of the womb, becomes attached to it, and starts to grow. The cells divide, doubling up each time, and an embryo forms. It goes through various stages of development, later becoming known as a foetus, and gets its nourishment from the mother's own blood supply, through the umbilical cord which connects it to the placenta. There, the blood supplies of the embryo and the mother meet, and food supplies and waste products are passed from one to the other.

Infertility is the condition when animals are not able to breed properly and produce offspring. An animal is sterile when it cannot breed at all. Infertility may be due to several reasons: damage to the sex organs; disease; overfeeding which makes animals too fat; poor diet, resulting in shortage of protein or of certain vitamins; wrong balance of minerals in the diet.

Male or Female Sex

The sex of an animal depends on chance. How the sex is decided is different in animals and birds, but it occurs at the time of fertilisation.

So far there is no means of selecting male or female in advance, although it is now possible, in some cases, to influence the choice of sex. This is a matter still under consideration.

The sexes are born nearly in equal numbers, but as it is by chance, there may be several males in a row, or a litter of pigs may be all female. These are the average figures:

	Male	*Female*
Pig	110	100
Cattle	105	100
Sheep	98	100
Horse	98	100

Numbers of Offspring Produced

The number of eggs released by the ovary varies with the type of animal. It is usually 1 or 2 in the cow at each heat period, 2 or 3 in the ewe, 20 or more in the sow. Some breeds produce more than others—these are known as more prolific.

Even when all the sow's eggs are fertilised, some of them fail to develop, particularly if the sow is not fed properly. An average number born in a litter is 12. If a sow is served more than once when she is mated, she is more likely to produce a larger litter.

Flushing means special feeding for ewes just before they are mated. Giving extra food, preferably rich in protein, for a few weeks before mating increases the chance of getting a larger crop of lambs. Such feeding encourages the ovaries to release more eggs.

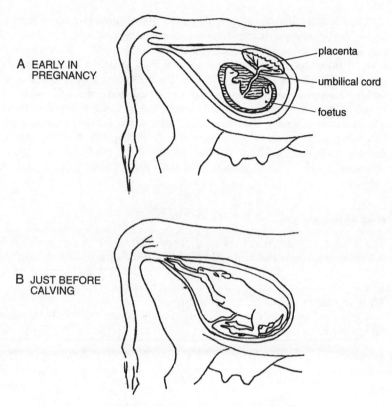

Figure 27   Development of young animal inside the mother

Ewes normally produce 1 or 2 lambs (twins in 45 per cent of cases) according to breed and management. Sometimes 3 lambs are born together, very seldom 4 or even more.

Cows usually produce a single calf. Twins are found in about 1 case in 80 calvings.

Twins   Ordinary twins occur when two separate eggs are fertilised at the same time. Identical twins are found when a single egg is fertilised and splits in two, each of which develops into an embryo. Identical twins are of the same sex and very much alike. They are particularly valuable for scientific research work.

Freemartins are found in calves, and are rather rare. When a heifer and bull calf are born twins, the heifer calf is known as a freemartin. When she gets older she will not be able to breed in 9 cases out of 10—so there is no point in keeping a freemartin for breeding.

This happens because both calves are fed by the same blood system, and the male hormones of the bull calf reach the heifer calf in this way and suppress her sexual development.

Pregnancy

When the egg is fertilised and starts to grow in the womb, the animal is pregnant. The embryo starts very small and grows slowly during the first two-thirds of pregnancy, then grows very quickly during the last third—when extra feeding is needed for the mother.

Common signs of pregnancy are:

- Oestrus stops—the animal no longer comes on heat.
- Milk yield drops.
- The animal is quiet, seems in good condition, and the body swells.
- The udder starts to develop ('bagging up') towards the end of pregnancy.

Pregnancy testing to determine whether female animals are pregnant can be done on the farm by staff or by specialist contractors. Various forms of pregnancy diagnosis (PD) are used, including ultrasonic and visual scanning for sheep, milk testing for cows, and the use of probes to check on blood flow; internal examination by a veterinary surgeon is also used.

Abortion means that the young creature is born too soon, usually at such an early stage that it does not have a chance to live. The main reasons for abortion are injury and damage to the mother or fright (such as caused by dogs attacking in-lamb ewes); disease (such as brucellosis—see page 116); poor feeding during pregnancy.

Extra feeding is needed by a pregnant female towards the end of pregnancy. The embryo is very small during the beginning of pregnancy, but grows very quickly towards the end.

Gestation

The gestation period is the length of time between mating and birth. The times that follow are only averages. If this period is longer in a cow, a bull calf is usual.

| | Gestation periods | |
	Days	Weeks
Sow	112 to 120	16
Cow	281 to 285	40
Ewe	140 to 147	20
Mare	340	48

Birth

Common signs that birth is near are:

- The animal is uneasy, often goes off on its own.
- The udder develops and swells up ('bagging up').
- The vulva swells, and there may be some discharge.
- The pin bones open up, and the muscles relax around the tail head.

The first thing to be produced is the water bag, which bursts and releases fluids which act as a lubricant. The mother strains and the young creature is forced out. The normal position of the newborn is with the head and the front feet first. Help is needed if birth is very slow, if any difficulties occur, or if the position is wrong. After the young creature is born, the afterbirth (cleansing) comes away; this is the placenta.

Care of the Young Creature

Make sure it can breathe, if necessary by clearing the nose and mouth. The mother will usually lick at it first, and it is sometimes as well to dry it by rubbing it firmly with a dry sack. The young animal is connected to its mother by a cord, which breaks; the navel and end of the cord should be disinfected to prevent germs causing infection.

Colostrum is the first milk of the mother after birth; the young creature should have this, as it is special milk—containing extra protein, minerals and also antibodies which protect against common infections.

SYSTEMS OF BREEDING

Species are the main groups of animals. We are concerned in British farming with cattle, sheep, pigs, horses, chickens, ducks and turkeys. These cannot be crossed with one another—breeding is kept within the species.

Breeds of these different animals and birds are well known. They usually differ from one another in appearance, and often in performance too. Most British breeds came into being during the nineteenth century. Breeds of animals are commonly crossed (mated) together.

Strains are found within the breeds. They may look alike but perform differently, or may even have a different body shape.

The main systems of breeding used by farmers are given below.

Crossbreeding

This means mating two different breeds together. There are various reasons why breeds are crossed in this way:

- to get some of the qualities of each parent in the offspring, and thus produce an animal which is more useful and/or more saleable
- to improve the quality of production from livestock, whether beef, milk or wool

Examples are:

BORDER LEICESTER RAM × SCOTTISH BLACKFACE EWE
(good carcase quality; prolific) ↓   (hardy; good mothering)
GREYFACE LAMBS
(good productive quality)

HEREFORD BULL   ×   FRIESIAN COW
(good beef quality)   ↓ (good milk production and growth)
HEREFORD / FRIESIAN HEIFERS
(suitable for breeding beef animals,
combining the characteristics of both breeds)

SUFFOLK RAM   ×   WELSH MOUNTAIN EWE
(good carcase quality)   ↓   (good mother)
CROSSBRED LAMBS
(for fat lamb production)

Hybrids are produced by crossing together two different breeds, or in some cases by crossing two strains or types. In the case of poultry and pigs today, a good deal of scientific work has been put into producing hybrid animals for specific purposes. An example is:

LARGE WHITE BOAR   ×   WELSH SOW
(from tested strain)   ↓   (from tested strain)
LARGE WHITE/WELSH GILTS
(with qualities of food conversion rate,
carcase quality and rate of growth of parents)

Hybrid vigour is 'the little bit extra' which comes from some crossbreeding when, for example, the production of the cross between two breeds exceeds the average of the production of the parents. Sometimes the young from a cross start off with more strength and vigour than purebreds from either parent.

Most poultry now are bred in this way, using mixtures of pure breeds on each side, to produce a bird with special qualities.

Grading-up is a process of crossing, using males of one breed on females of another breed for 4 or more generations. By this time, the original breed of the females has practically disappeared, and the breed

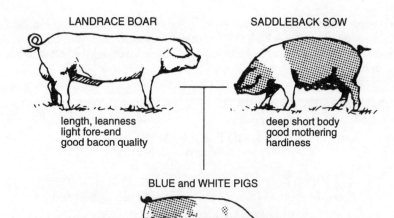

LANDRACE BOAR

length, leanness
light fore-end
good bacon quality

SADDLEBACK SOW

deep short body
good mothering
hardiness

BLUE and WHITE PIGS

fair length
not too deep and short
good grower
plus
hybrid vigour

Figure 28   Commercial crossbreeding

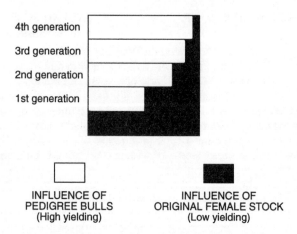

4th generation
3rd generation
2nd generation
1st generation

INFLUENCE OF
PEDIGREE BULLS
(High yielding)

INFLUENCE OF
ORIGINAL FEMALE STOCK
(Low yielding)

Figure 29   Grading-up dairy cows

of the males has taken its place. For example, many herds of cows have become Friesian in recent years, starting off from other breeds and using Friesian bulls on them. In the same way, farmers will cross Large White boars on other breeds of pigs until all their pigs become Large White.

Grading-up is also used within breeds, to make a herd all-pedigree, by the use of pedigree bulls on 'foundation cows' of a good type. This is allowed by most dairy breeds, but not by some beef breeds.

Pure Breeding

Normal pure breeding is mating purebred animals of the same breed— for example, Friesian bull and cows, Large White boar and sows, Clun ram and ewes.

The use of good quality testing makes it possible to improve the quality and the production of a herd or flock.

Line breeding is the use of a family of males on a herd or flock. Starting with one bull, you would change to one of his sons when there was a danger of the crossing becoming too close. This allows grandfather to be mated to granddaughter, but not to his own daughters.

In-breeding is the use of very close crosses, such as sire to his own daughters, son to mother, or brother to sisters. It is a breeding practice only done by very skilled breeders who know exactly what they are doing; it is not a practice for the ordinary farmer. In-breeding can pass on some of the best qualities in stock, but also identifies bad (and even harmful) qualities.

Normal Breeding Practices

Normal breeding practices with farm livestock today are:

Dairy cattle Mostly pure breeding by the use of AI. High quality, high production bulls are available at AI centres, and in time are gradually improving performances in our dairy herds. Farmers can 'nominate' the use of a selected bull by paying an extra fee.

Beef cattle Mostly produced by cross-breeding, using beef bulls (by AI or natural service) on some dairy cows, particularly the lower yielders. Bulls used in this way are Hereford, Charolais, Aberdeen Angus and North Devon, which 'colour mark' their calves (see page 103) because of a dominant character.

Sheep A great deal of crossbreeding is done, much via the use of quality rams on hardy ewes to produce lambs which will grow well on better conditions than their mothers are suited to. Thus lambs are produced from hill and mountain flocks and brought down on to easier and better conditions, where they make good growth. Down rams are used on most other breeds for finished lamb production.

Pigs There is a good deal of pure breeding, using the main white quality breeds (Landrace and Large White), and also crossbreeding, using a white quality boar on other sows to produce pigs which grow and feed well. The main white breeds are often crossed together. There are now a number of specialist hybrid pigs.

Poultry Most of our poultry today are specially produced, scientifically designed hybrids, suited to one particular purpose.

LIVESTOCK IMPROVEMENT

A great deal of work has been done to improve the quality of British livestock, so much so that this country has been called 'the stud farm of the world'—supplying breeding stock to many other countries. Some British breeds have been used very widely both as pure breeds and to improve the quality of native stock—Shorthorn, Hereford, Angus and Devon cattle, Southdown, Kent and Leicester sheep, Large White and other breeds of pigs.

To start with, British livestock was very mixed, with many local types, and with quality that varied from poor to indifferent. Improvement really started in the eighteenth century.

The Great Breeders were men who did wonderful work in improving certain breeds from poor, untested material. They included Robert Bakewell (Leicester sheep and Longhorn cattle), John Ellman (Southdown sheep) and Joseph Tuley (Large White pigs).

Agricultural shows gave breeders and farmers a chance to compare livestock and to see improved types and new methods. Local shows (such as the county shows) helped in this way, as did the big national shows. The first Smithfield Show was held in 1798 and the first Dairy Show in 1876.

Breed societies came into being to look after records of pedigree livestock. Their records were kept in the herdbooks and flockbooks. It is only by having proper records of this type that real progress can be made. Most breed societies were started in the second half of the nineteenth century.

Artificial insemination (AI) and embryo transfer services for cattle are largely carried out by the Genus business (details from Genus, Westmere Drive, Crewe CW1 1ZD) and by some other firms for pigs. AI has had a tremendous effect on breeds and breeding practices in recent years and has influenced livestock breeding very much.

AI is now a common farm practice with cattle and is becoming more common with pigs. Semen is taken from bulls which are kept at special centres. The semen is mixed with preservatives and may be stored for a time. It is put into cows on heat by specially trained inseminators. AI has several advantages:

- Fewer male animals are needed, which saves trouble and expense.
- It is not necessary to keep a male animal on each farm.
- Better quality sires can be used, which would otherwise not be within the reach of smaller farmers.
- There is a wide choice of breeds, and farmers can nominate a particular bull (ask for its semen to be used, for which an extra fee is charged).

Testing is essential in any scientific breeding programme. There are two chief methods:

- *Performance testing* means testing how an animal performs, whether in milk yield, milk quality, rate of growth or carcase quality. This gives useful information, but it does not always mean that the particular animal will pass these qualities on in the breeding.
- To know this needs *progeny testing* which is a test of just which characters and how many of them are passed on to the offspring. Studying the milk records of daughters and comparing them with the records of the parents shows the parents' worth for breeding purposes. In the same way, pigs can be tested by comparing the growth and quality of their offspring. There are various progeny testing stations for this purpose.

THE SCIENCE OF BREEDING

Behind all breeding practices and policies is the science of breeding which is known as genetics. It is a very complicated subject, and it is not an exact science. It deals not with the results of single animals but with the average results and performances of groups of animals. Some of the rules of genetics were first discovered by the Austrian, Gregor Mendel, during the early nineteenth century.

As the science of breeding is better understood today, it is possible for important work in breeding to be done. Some of this is in the hands of private breeders—the pedigree men who own well-known herds and flocks—and the breed societies. But much breeding work needs records of vast numbers of animals, scientific staff and scientific apparatus; all this is expensive, and this sort of work is in the hands of organisations such as the Milk Marketing Board, research stations and large private firms.

There are a few basic principles of breeding which we should understand so far as they affect farm livestock.

Characters behave as separate units in inheritance. Each individual animal is made up of a large number of characters. Some of these we can see: coat colour, shape of horns, pattern of markings, the white face of the Hereford. Some cannot be seen: milk yield, rate of growth, health and vigour.

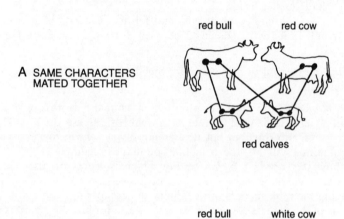

**A SAME CHARACTERS
MATED TOGETHER**

red bull   red cow

red calves

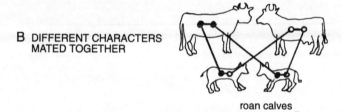

**B DIFFERENT CHARACTERS
MATED TOGETHER**

red bull   white cow

roan calves

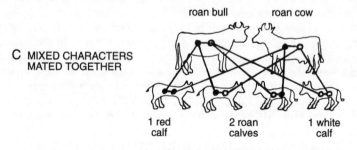

**C MIXED CHARACTERS
MATED TOGETHER**

roan bull   roan cow

1 red
calf

2 roan
calves

1 white
calf

Figure 30   Principles of breeding with no dominant character

Example: Cross a Hereford Bull and a Friesian cow. The Hereford gives a white face, red coat, low milk yield. The Friesian gives black and white coat, high milk yield. When these are put together the calf produced will have a white face, a black coat, and milk yield in-between the two. The characters have sorted out separately.

When animals with two different characters are mated together, in the first generation the offspring will show a mixture of these characters. In the second generation, the characters are sorted out and separate among the offspring.

Example: Mate together a Red Shorthorn and a White Shorthorn. The calves will be a mixed red and white colour called roan. If we mate two of these roans together, the offspring may come in any order, but on average any four of them will be sorted out this way: one red (like the grandfather); one white (like the grandmother); and two roan (the mixture).

Remember: the proportion is 1:2:1 or in the words of the old rhyme: 'One White, One Black and Two Khaki'.

Some characters are dominant to others in breeding. While other characters are really there in the offspring, only the dominant characters show. The characters which do not show—which are weaker—are called recessive.

A Dominant Character in Breeding

Common dominant characters are:

In cattle
 Black coat is dominant to white coat.
 Single-colour coat to spotted or patterned coat.
 Polled (hornless) to horned.
 White Hereford face to other types of face.

In pigs
 Black coat to red coat.
 White to other colours (not always completely).

If we cross Aberdeen Angus (black, polled) and Red Poll we get all-black calves. They are really a mixture of characters, but only the black shows; the red is there, but we cannot see it. If we cross together two of these animals, there will be a sorting out in this second generation in this way: one black (like the grandfather), one red (like the grandmother), two black (but a mixture, carrying both characters, like the parents). The proportion is really 1:2:1 but because one colour (black) is dominant, the offspring appear to be three black to one red.

If we mate together a Hereford (red with a white head) and an

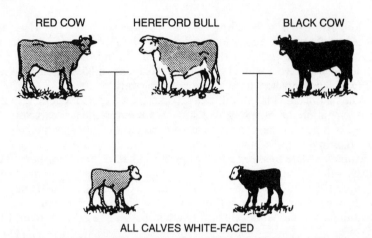

ALL CALVES WHITE-FACED

Figure 31   A dominant character in breeding

Aberdeen Angus, there are three dominant characters—black colour, a polled head, the white Hereford face. The calf will have all these, and so will be all black in body with a white face and no horns.

Understanding these principles of genetics does make a few points clear:

- An animal is not always what it appears to be. It may be pure for a particular character (such as colour of coat) or it may be mixed.
- In breeding, an animal will not always behave according to how it looks. It may have a mixture of characters rather than pure characters and these are not always passed on to the offspring in a simple way.
- In crossbreeding, where any character is dominant, there is always a recessive character somewhere, and this may come out in the breeding. Examples of this are Red Friesians or Red Angus which sometimes appear—because two animals with these red recessive characters have been bred together.
- Offspring are not always like their parents; they may turn out more like their grandparents.

Some important characters which affect the profit of an animal—size, growth, food conversion, etc—are due partly to its breeding and partly to the effect on it of management and feeding. It is therefore worth while breeding for some characters, and a waste of time (or else a very long job) breeding for others. A low percentage means that breeding does not influence the character much; a high percentage that it does and breeding for is worthwhile.

Character	Percentage heritability
Litter size	10%
Birth weight	15%
Growth rate	30%
Food conversion	35%
Backfat thickness	50%
Percentage lean meat	50%
Length of back	70%

GROWTH OF ANIMALS

All animals grow in certain stages, and in each of these stages there is a different type of growth. This is not always obvious, particularly in early finishing systems where the animal goes for slaughter before it has finished growing. There are three stages of growth:

1. Frame The young animal is all skin and bones; its legs and head are out of proportion. As it develops, its bones grow to provide the frame which supports the whole body later. Its internal organs grow to provide the digestive system and all the other systems of the body.

2. Flesh (muscle) While still growing, the animal forms muscle. The body becomes longer and deeper, and it puts on weight quickly. The meat (muscle) is young, tender and light-coloured at this stage, without much flavour—and it is this we get from the veal calf, the broiler chicken, the pork pig and the young lamb.

3. Fat Towards the end of the period of muscle development, the animal starts to fatten. Fat is stored around the outside of the body under the skin, and later it is stored around the kidneys and other internal organs and in such places as the scrotum and around the tail-root—which shows that the animal is getting 'finished'. Older animals also store thin layers of fat inside the muscle; in beef cattle this is known as 'marbling'.

Early fattening and quick growth run these stages together—telescope them—so that the young animal is putting on fat and muscle together, even before it has finished making its frame.

Early developing types start to put on fat younger. *Example*: the short fat types of pig, which tend to get too fat at bacon weights.

Late developing types do not put on much fat until they are older. *Example*: the long bacon type of pig, and Friesian cattle.

Remember the order: FRAME—FLESH—FAT.

THINGS TO DO

1. Get to know the signs of heat, pregnancy, and approaching birth of farm animals.
2. Learn how to keep service, calving, litter and other records of livestock.
3. Weigh young animals at birth and at the various important stages of growth.
4. Estimate the weights of all classes of livestock and check these against the weighing machine.
5. Learn the appearance of the different important breeds and crosses of farm livestock, and look for dominant characters in breeding.
6. Get to know and understand milk records, pig records and AI and pedigree records.
7. Learn how to contact AI services and what to do in preparation for the inseminator.

QUESTIONS

1. What happens during the processes of mating, fertilisation, and the period up to the birth of the young animal?
2. What are the gestation periods of the main farm livestock—cattle, sheep and pigs?
3. Why is crossbreeding carried out? What benefit can it produce in the offspring of a cross?
4. What are dominant characters in the breeding of farm livestock? What is their practical importance on the farm?
5. What are the stages of development in the growth of a young animal? How does this development affect meat production?

Chapter 6

Animal Health and Disease

THE HEALTH of farm livestock depends partly on good management, and partly on the knowledge of the farmer and stockman about the dangers and risks of disease and other troubles, with the guidance and help of the veterinary surgeon.

ANIMAL HEALTH

All farm livestock should be kept in good health. If they are not, they cannot do their job properly, whatever it is. An ill animal is unhappy, uncomfortable and usually loses its appetite. This condition interferes with growth and production. A sick animal is no longer profitable.

What the Animal Needs to Keep it in Good Condition

Good feeding The animal needs food suited to its digestive system and to its age and condition. It needs enough food to cover all its needs—energy, warmth and production. Its food must be properly balanced—with enough energy, protein, fibre, vitamins and minerals. It must have enough water.

Good housing Some animals, such as sheep and some beef cattle, are out all the time and do not normally need housing. But most farm stock are housed at some time. Pigs, poultry and calves need to be very well housed.

Healthy conditions On the land, livestock must not be kept on the same field or paddock too long (year after year) or crowded too closely together. The old saying 'a sheep's worst enemy is another sheep' means that where too many are kept together, germs and parasites pass from one to another and build up until there is trouble. A change of land is important, and young animals should go onto clean land. The same applies to buildings, which should be rested, cleaned and disinfected between batches of animals.

107

Protection from disease   Every normal precaution must be taken to protect animals from germs and other causes of disease. If fresh animals are brought onto a farm, keep them isolated from the others for a time. Beware of animals which have moved about the country too much, or been in and out of markets. Isolate any stock which look at all ill, so that they do not infect the rest.

SIGNS OF HEALTH

The normal signs of good health are listed earlier in this book under each section of livestock.

An animal's temperature is a good indication of its state of health. It is usually taken by putting a veterinary thermometer into the rectum for a few minutes; normal temperatures are shown below.

	Body temperature °C	°F	Pulse rate (per minute)	Breathing rate (per minute)
Pig	40	104	70–80	10–16
Cow	39	102	45–50	12–16
Sheep	40	104	70–80	12–20
Chicken	41	106	120–160	15–45

(Note: these are average normal figures. There is a good deal of variation according to age, condition, size, etc.)

The following signs apply to all livestock: a healthy animal eats well, produces dung and urine in a normal condition, walks and holds itself properly, looks lively and contented.

Causes of Ill Health

Some animals are 'poor doers'—they grow badly, never have much vigour, suffer from various bad conditions and are more likely to catch diseases than the rest. A 'runt' pig is an example of this; it starts late and never catches up. Personal attention and special care can help such an animal, but it is not always worth it. Apart from these cases, the main causes of ill health are:

Mistakes in management   Where there is something wrong with feeding, housing or hygiene, trouble can be expected. This is seen commonly, and very quickly, with intensive poultry. Examples of these troubles are mineral and vitamin deficiencies, chills caused by cold and damp, and ringworm or lice caused by unhygienic housing. Stress is a word often used today to describe many of these causes of trouble.

Infection There is little or no protection against some of the very infectious diseases; if the germs are introduced to a farm or a herd they may cause widespread trouble. Examples of this are foot and mouth disease, anthrax and swine fever, all notifiable diseases (see page 111).

Lack of protection Where trouble can be expected with some diseases, it is wise to protect stock by inoculating them. If this is not done, you can only blame yourself if trouble follows. Examples are pulpy kidney disease in lambs and erysipelas in pigs.

TYPES OF DISEASE

If we take the word disease to mean any trouble or abnormal condition of farm livestock, there are several groups of diseases, with different causes.

Disease may be mild (like indigestion) or serious (like anthrax, a killer). It may be chronic, lasting a long time (like some forms of mastitis), or acute, coming on suddenly (like foot and mouth disease).

These are the main types of disease:

Infection by organisms (germs) There are many different bacteria found naturally everywhere; most of them are harmless, but some cause disease. Bacteria are very small forms of life which multiply very quickly. Examples of disease caused by harmful bacteria are mastitis, brucellosis, tuberculosis, lamb dysentery.

A virus is a living chemical, and some viruses cause disease. Examples: virus pneumonia, foot and mouth disease, swine fever.

Parasites These are living things, plant or animal, which feed on other living things and cause harm to them. There are many parasites which live off farm animals and these can be divided into two main groups.

- *External* parasites live outside the animal, usually feeding on its skin or hair. Examples: ringworm (a fungus), lice, ticks, maggots (insects).
- *Internal* parasites live inside the animal, usually in the digestive system. Examples: liver fluke, lungworms, stomach and intestinal worms, warble fly larvae.

Nutritional deficiencies Shortage of something in the food. Deficiency of some mineral or vitamin can cause trouble, and so also can a wrong balance of minerals. Examples: piglet anaemia (iron), rickets (vitamin D), milk fever (calcium).

Metabolic disorders Some upset in the animal which may be due to stress and strain, wrong management or a variety of causes. Examples: bloat, milk fever, acetonaemia, twin lamb disease.

Poisoning Certain plants and poisons affect livestock, may cause upsets or may kill them. Examples: lead poisoning, yew trees, weeds such as ragwort or horsetail.

CONTROL OF DISEASE

Management has already been covered (see page 108). It includes good feeding, proper housing and clean and healthy conditions.

Hygiene covers all the commonsense precautions that should be taken to prevent disease from coming on to a farm or spreading when it is there. It includes clean land, clean buildings, quarantine (keeping separate) of bought-in animals and taking care with visitors who might bring infection (dip your boots in disinfectant, please).

Stockmanship is important in two ways. A good stockman practises observation and sees the first signs of trouble or illness. He should act at once. When disease is found he must do whatever is necessary; in many cases this means carrying out the veterinary surgeon's instructions. In this case, the good stockman acts as an animal nurse.

Medicine includes all the chemical means by which disease is treated. Some simple medicines are kept on every farm and are used by the stockman—such things as udder salves for sore teats, antiseptic ointment for sores and wounds, drenches for indigestion, worming medicines. Other more potent medicines are only used at the direction of the vet—these include antibiotics and other drugs.

Veterinary aid must be called in for many troubles, and the sooner the better. A difficult birth, wounds and accidents, a bad case of bloat, and any disease other than the mild and minor cases need a visit from the vet.

Inoculation (usually by injection) is often done by the vet but sometimes by the farmer or stockman, as in the case of sheep. To protect against a number of diseases, either a vaccine or a serum may be injected.

A *vaccine* is a prepared substance which is injected into a healthy animal: this makes the animal produce antibodies which protect it against a particular disease. It builds up resistance to the disease, and is immune to it, but this takes some time to develop, maybe 2–3 weeks. This immunity can be passed on to the offspring through the colostrum in same cases—the colostrum contains antibodies.

A *serum* is prepared from the blood of an animal which has been specially exposed to a disease. Injecting a serum into a healthy animal gives immediate protection—the animal becomes immune to the disease at once—but this protection may not last very long and may have to be repeated.

There is also natural immunity. Some animals are naturally resistant to certain diseases, which may be due to the protection given by antibodies in the colostrum they got in their first few days. It may also be due to exposure to small doses of infection or parasites, which allows the animal to build up its own resistance.

Homeopathy This system of health care has been in use for humans and also for animals for some years, and it is now becoming better known. It is practised by a number of vets who claim success, often at low cost.

In homeopathic treatment, very small doses of specially prepared medicines are given, and these work on the whole system of the animal. Individuals or even whole herds or flocks can be treated.

Careful observation by the person in charge of the animals is necessary. The system is described in more detail in a number of specialist books.

The law The farmer and the stockman have legal duties. If they suspect one of the *notifiable diseases*, the police or a Ministry veterinary officer must be told at once and any infected animals isolated. After this it becomes the responsibility of Ministry of Agriculture vets, who will tell you what to do.

The notifiable diseases found in Great Britain now include:

Anthrax, Aujeszky's Disease, Blue Tongue, Bovine Spongiform Encephalopathy (BSE), Cattle Tuberculosis (certain forms), Foot and Mouth Disease, Fowl Pest, Rabies, Scrapie, Swine Fever, Warble Fly.

There are other notifiable diseases which are no longer found in this country or which are unusual. 'New' diseases can appear from time to time.

Make sure you know which diseases are common and which are notifiable. It can happen on your farm.

A record of the movement of animals must be kept, so that any spread of infection can be checked accurately: this is done in a special book, which must be available for inspection at any time.

Apart from the notifiable diseases, there is government control of animal health in order to stamp out certain diseases which once were widespread. Testing, slaughter policies and vaccination have been used to control tuberculosis, brucellosis and fowl pest.

DISEASES OF PIGS

Notifiable Diseases

Foot and mouth disease is caused by a virus. It affects all cloven-hoofed animals (cattle, pigs, sheep). The official policy in Great Britain is to slaughter all infected animals and others that have been in contact with them, and so to keep the disease under control.

Symptoms Lameness is the chief sign in pigs. There is also pain, the pig does not like being moved, high temperature. Blisters on feet, snout, udder, teats, mouth. Loss of appetite, scouring or constipation.

The disease is spread by diseased animals, litter, lorries, persons, sacks, foodstuffs, birds, frozen carcases, and swill which has not been boiled properly.

Precautions and control
- Infected animals and contacts are slaughtered
- Movement restrictions
- Disinfection of all premises
- Boiling of swill

Swine fever This disease, which is caused by a virus, was eradicated in Britain, but there have been further outbreaks. It takes various forms, both acute and chronic. Swine fever is spread by infected animals, feedingstuffs, lorries, utensils, etc. It causes high temperature, abortion in sows, loss of appetite, and often death in 2–4 days. The disease is controlled by a slaughter policy.

Aujeszky's disease A disease caused by virus infection which has become established in intensive pig farming areas. It is most serious in young pigs and often leads to death.

Anthrax This affects all farm animals and also humans. It is caused by bacteria, and is very infectious. It is an acute disease, usually fatal.

Other Diseases

Swine erysipelas is caused by bacteria. It affects pigs of all ages, but larger ones mainly, and usually occurs in hot or muggy weather. It is commonly called 'Diamonds' because of the red skin marking it causes.

Symptoms Inflammation and skin pustules—red markings of the skin—diamond shaped; constipation; loss of appetite; lameness—joints swollen, usually in pigs of bacon weight or upwards; death sometimes within 24 hours. It is spread by a germ which is very common in all places where pigs are kept.

Precautions and control
- Vaccinate breeding stock annually with swine erysipelas serum.
- Isolate and/or vaccinate new stock coming to farm.

Virus pneumonia is caused by a virus and affects many herds of pigs, causing general poor health and performance.

Symptoms Erratic, hard cough, especially when aroused. Pigs are unthrifty and they may take much longer to fatten. It is spread by virus coughed into the air and direct contact from the sow to her litter.

Precautions and control
- Buy breeding stock from a virus-pneumonia-free herd
- General cleanliness
- Fresh air
- Disinfect pens
- Farrowing and rearing in isolation

Enteritis is another very common disease. One form (transmissible gastro-enteritis—TGE) is caused by a virus. It mainly affects very young pigs, and is very infectious.

Symptoms Diarrhoea and inflammation of the bowel.

Precautions and control This must be treated as a farm problem; consult the vet.

Anaemia is caused by a deficiency of iron in the blood. It will affect young pigs (up to 3 weeks old) if not prevented and is caused by a shortage of iron in the sow's milk.

Symptoms With young pigs—skin pale, ears yellowy, diarrhoea dirty grey or yellow. Tenth day critical. Breathing difficult. Loss of condition or death.

Precautions and control
- Little pigs should be given an iron preparation by injection or by dosing.
- Pigs running on grassland are not so likely to suffer.

Parasites

- *External* Found on the skin of the pig; there are two of chief importance.

Lice are grey insects, about 6 mm ($\frac{1}{4}$ in) long. They do little harm unless they are found in large numbers, when they cause unthriftiness and irritation.

Treatment Dust or soak the skin with treatment which kills the lice.

Mange is caused by very small mites which burrow under the skin, producing sore parts. The skin becomes rough and wrinkled, and the pigs unthrifty; they are always uncomfortable and rub and scratch continually.

Treatment
- Scrub with soap and water and apply a treatment which kills the mites.
- Sows are washed before going into their farrowing quarters.

- *Internal* Inside the pig's body; there are two types of worm.

Roundworms live in the small intestine of pigs and are very common. The large roundworm is as thick as a pencil and up to 30 cm (12 in) long. The females lay eggs inside the pig and these eggs pass out in the dung on the ground and can contaminate food. Moisture makes them ready to hatch out in 3 weeks or more, according to weather conditions. The eggs are swallowed and the little larvae hatch out from the eggs inside the pig's intestines, pass through the body to the lungs, climb up the wind pipe and are swallowed. In the intestine the worms become full-grown and the females start laying eggs.

Symptoms   The pigs do badly, cough, suffer from digestive troubles, and may go yellow with an attack of jaundice.

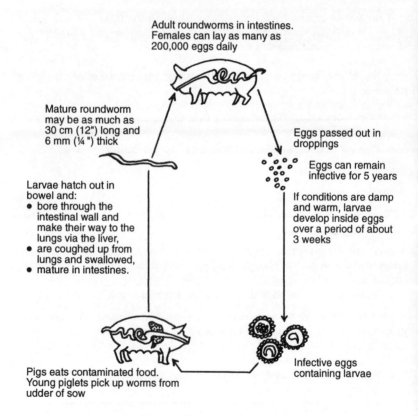

Figure 32   Life cycle of the pig roundworm

Precautions and control
- Avoid keeping pigs too long on one piece of land—move them onto clean land.
- Thoroughly wash and disinfect pens and buildings between batches of pigs.
- Give sows worming medicine and wash their udders thoroughly before farrowing.
- Give worming medicine to all pigs.

Lungworms are white thread-like worms, 25–50 mm (1 to 2 in) long. They live in the tubes of the lungs where the females lay eggs. These are coughed up and swallowed, and the eggs pass out in the dung. Some of these eggs are eaten by earthworms, and the larvae of the lungworms develop inside them. Pigs eat earthworms, and the larvae pass from the pig's intestine to the lungs, where they develop.

Symptoms Coughing, poor condition, pneumonia and death in bad cases.

Precautions and control
- Run pigs on clean pasture and never keep them too long on any piece of land.
- Pigs kept in buildings do not suffer from lungworms.

DISEASES OF CATTLE

Notifiable Diseases

Bovine spongiform encephalopathy (BSE)   BSE, sometimes called mad cow disease, was made a notifiable disease in 1988. There is no treatment, no cure. It occurs in adult cattle as a disease of the brain, and all suspected cases must be reported to the Ministry of Agriculture. There is a strong suggestion that this disease is somehow related to scrapie in sheep.

Foot and mouth disease
Symptoms Rise in temperature, the animal is dull, blows slightly, goes off its food, stops chewing the cud. It moves its lips, sucking and dribbling, and there is pain in the feet and mouth.

Anthrax is caused by bacteria and is very infectious. It is an acute disease, usually fatal.

Symptoms High temperature and other signs of sudden illness; blood from the nostrils and in the dung; sudden death.

There is great danger from an infected animal or dead body. Do not cut it or mess about with it in any way. Keep away from it.

Tuberculosis is caused by bacteria. Not all forms of TB are notifiable, but if there is TB of the udder, a chronic cough, or an animal very thin and wasting away, the authorities must be told.

Tuberculosis in cattle has been largely eradicated in this country by a policy of 'tuberculin testing' and the slaughter of animals which react to this test.

Miscellaneous Troubles

Mastitis is the commonest disease of dairy cows. An inflammation of the udder, caused by bacteria, often started by dirty methods, bad management or damage. Mastitis is contagious in some forms.

It can be either:
- *Streptococcal*   The commonest type

Symptoms   Clots in milk (use strip cup to test milk), later the milk changes from normal and becomes thick. Parts of the udder may be hot and swollen. Milk yield drops (inside of the udder is destroyed).
- *Staphylococcal and summer mastitis*   These are less common and are more acute types. Summer mastitis is very acute as a rule and affects dry cows and heifers during July–September.

Symptoms   Udder becomes swollen, hot, painful. Part or all of udder may be destroyed. High temperature, fever, sometimes death.

Precautions and control of all mastitis
- Cleanliness, good milking routine
- Good management
- Use of disinfectants and teat dips
- Treatment with antibiotics (as advised by the vet)
- Inspect dry stock. Control flies

Brucellosis (contagious abortion) was previously a widespread disease of cattle. Caused by bacteria and easily spread, it used to cause serious losses. An official government eradication scheme was successful, although isolated cases continue to be reported.

Symptoms   Abortion—usually in first or second pregnancy. Cows often abort once and then no more, but pass on infection afterwards. Inflammation of womb—discharges and blood. It is spread by infected food, water or surroundings, germs entering the body through the skin, etc.

Virus pneumonia is a common complaint of calves, mainly of 2 to 4 months old. It is more likely to occur where many calves are kept together, badly ventilated, with poor conditions and poor management.

Symptoms   Shivering, rise in temperature, loss of appetite, shallow and sometimes noisy breathing, and death. It is spread by coughing and sneezing, and encouraged by high humidity.

Precautions and control
- Airy, well-ventilated houses
- Small groups of calves with solid partitions between groups
- Isolation of new groups of animals
- The use of a vaccine

Scours is a very common disease of calves, caused by bacteria, and encouraged by wrong feeding, bad management, chills and too much travelling.
Symptoms Thin, watery, yellow, stinking dung, pain in the belly, 'tucked up' appearance. Calves waste away and may die.
Precautions and control
- Good management. Cleanliness.
- Cut milk out or reduce quantity, for a time. Feed boiled water and glucose.
- Make sure calf gets colostrum after the birth.
- Treatment with drugs.

Foul-in-the-foot is a complaint of cattle, particularly common in the winter under muddy conditions. It is caused by bacteria which get into the foot and cause inflammation. The use of a footbath containing 10% formalin helps to prevent this trouble. Bad cases need help from the vet, who will probably give an injection.

Infertility (animals unable to breed) can be very troublesome with cattle, particularly dairy cows. It may be caused by disease (contagious abortion, vibrio foetus, etc.); various forms of infection; lack or faulty balance of minerals in the diet.

Other Troubles not Caused by Infection

Bloat The rumen (first stomach) is blown up with gas, which causes great pain and may kill the animal. It is more common in some fields and with certain animals, but may be expected on very lush pasture, particularly in the spring.
Symptoms The belly is blown out, chiefly on left side, the animal becomes very uneasy, quick distressed breathing and pain.
Precautions and control
- Be very careful on lush pasture containing much clover. Graze only for a short time.
- Feed hay or straw before grazing.
- Walk animals around, it not too far gone.
- Drench with oil or special preparations. Put a tube down the throat into the rumen.
- In bad cases get a veterinary surgeon as soon as possible. It is possible to release the gas by puncturing the rumen.

Ketosis (acetonaemia) is found in high-yielding, highly fed cows, within a few weeks after calving.

Symptoms Loss of condition. Sweet smell (acetone) in breath, urine and milk.

Precautions and control

- Avoid overfeeding before calving.
- Do not reduce roughage in the diet too much; feed some hay daily. Make sure a balanced high energy ration is fed, particularly in the first two weeks after calving.
- Good careful management.

Milk fever A complaint of high-yielding cows after calving. It is not a fever and is caused by a shortage of calcium in the bloodstream.

Symptoms Usually found within 3 days of calving. Uneasiness, swaying, animal drops down, may go into coma, head twists around, panting, groaning, cow may die.

Precautions and control

- Make sure there is plenty of calcium in the ration.
- Prop the cow up with bales of straw.
- Get the vet at once.
- Injection with calcium preparation usually cures the complaint fairly quickly.

Hypomagnesaemia (grass staggers, grass tetany) is caused by a shortage of magnesium in the bloodstream. It is found in cattle grazing on certain pastures and is commoner in some districts. It is now common in dairy herds, and although it may be found at any time of the year is worst in spring after the cows are turned out.

Symptoms Nervousness, loss of appetite, milk yield may drop suddenly, animals may drop down, and die quickly.

Precautions and control

- If trouble is expected, feed powdered Magnesite daily or special magnesium rich foods.
- Feed properly and provide shelter for out-wintered stock.
- It is possible to apply special fertilisers or forms of lime (containing magnesium) to pastures.
- In acute cases, get the vet, who will inject a magnesium preparation.

Parasites

- *External*

Lice and mange can be controlled by scrubbing and the use of insecticides. Good feeding may also help.

Ringworm is caused by a fungus. It spreads to other cattle and to human beings and may live in the woodwork of buildings. It is not easy to treat, but various dressings can be used and medicines given.

• *Internal*

Warble fly   Eggs are laid on the legs or belly of cattle. The larvae burrow into the body and travel through it for 9 months or so, to the back of the animal. There they develop below the skin until they leave the body, fall to the ground and change into a fly.

Control   Farmers are encouraged to treat all cattle each autumn, as we move towards eradicating this pest. Warbles in cattle are now treated as a notifiable disease.

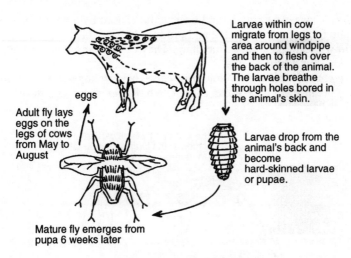

Larvae within cow migrate from legs to area around windpipe and then to flesh over the back of the animal. The larvae breathe through holes bored in the animal's skin.

eggs

Adult fly lays eggs on the legs of cows from May to August

Larvae drop from the animal's back and become hard-skinned larvae or pupae.

Mature fly emerges from pupa 6 weeks later

Figure 33   Life cycle of warble fly

Worms in stomach and intestines   There are several different types which live and breed inside cattle, laying eggs which pass out in dung, develop into little larvae which get onto the grass and then infect the cattle again.

Symptoms   Scouring, dirty tails and hindquarters, animals 'doing badly'.

Precautions and control
- Graze stock on clean land.
- Take particular care with young cattle; avoid overcrowding.
- Dose or inject stock with suitable medicines.

Liver fluke   Fluke (flat worms) live in the cattle's liver and produce eggs which pass out in the dung. The little flukes pass into a small snail found in damp land, and then pass onto the grass, where they are taken in by the cattle. See Figure 36.

Symptoms Animals in poor condition, becoming thin and miserable. Sometimes there is no sign, but the liver is spoiled, which reduces the value of the carcase.

Precautions and control
- Drain wet land.
- Control the snail by use of chemicals.
- Fence off wet places.
- Dose cattle with suitable medicines.

Husk A severe cough (and bronchitis) caused by a lungworm. Eggs are laid in the lung, hatch out and produce larvae which are coughed up, swallowed and pass out in the dung. They develop on grassland and are picked up by grazing cattle.

Symptoms Commonly found in calves and young stock. Coughing, heavy breathing, distress, can cause death. Rapid loss of condition.

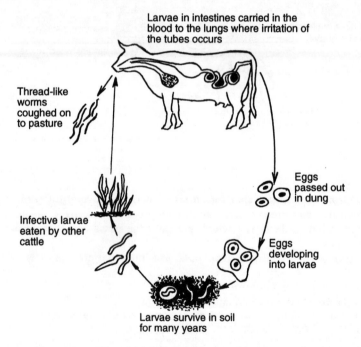

Larvae in intestines carried in the blood to the lungs where irritation of the tubes occurs

Thread-like worms coughed on to pasture

Eggs passed out in dung

Infective larvae eaten by other cattle

Eggs developing into larvae

Larvae survive in soil for many years

Figure 34   Life cycle of the lungworm which causes husk

Precautions and control
- Keep calves housed during early spring, autumn and winter.
- Mixed stocking of grassland.
- Treatment with drugs or vaccination.

DISEASES OF SHEEP

Notifiable diseases

These include *foot and mouth disease* (see page 115) and *anthrax* (see page 115).

Sheep scab is very easily spread. It is caused by mites which live on the skin of the sheep, causing scabs which irritate the sheep very much. To protect against this disease sheep are dipped. This dipping (or other treatment) also gives protection against fly (see page 123) and other skin parasites.

Other Diseases

Footrot is caused by bacteria which get into cracks in the feet. It is the commonest disease of sheep, found in nearly every flock, causes a lot of trouble which is not always realised, and can very quickly bring sheep down in condition. It spreads from infected sheep's feet through the soil, although the germ cannot live away from sheep for more than 2 weeks.

Symptoms Lameness, cracks and holes in the hoof, inflammation of the feet. A strong smell from the feet (very like ripe cheese).

Precautions and control
- Keep feet trimmed with a knife and/or clippers and sheep exercised to keep feet in good condition.
- Use the foot bath regularly—with a solution of formalin (10 per cent), zinc sulphate or other antiseptic.
- Isolate any infected sheep and treat them until footrot is controlled. Vaccines to help protect sheep against various strains of footrot are injected once or twice each year. Antibiotics should be kept for the worst cases.

Lamb dysentery is caused by very common bacteria. It is found on many farms, and is more likely where sheep have been kept for some time. It is very easily passed on through a flock, and trouble is common in young lambs.

Symptoms Found during first or second week after birth. Lamb is dull, lags behind ewe and it will not suck. Scour, sometimes with blood. Belly blown up and painful.

Precautions and control

There is no treatment for infected lambs; they nearly always die.

- Routine vaccinations of ewes, once before mating, and again shortly before lambing. Immunity is then passed on to lambs in the colostrum.
- Inoculate newly born lambs with lamb dysentery serum.

Pulpy kidney disease is caused by another strain of the bacteria which cause lamb dysentery. It is found in older lambs and can cause many losses in a flock.

Symptoms Found at 6–8 weeks old; sometimes with older lambs in the autumn. Best lambs are found dead. Lambs very dull, quickly go down and die.

Precautions and control

- Avoid any sudden change of diet.
- Watch lambs doing well on very rich food.
- Routine vaccination of ewes passes on immunity to the lambs in the colostrum.
- Serum can be used, to give immediate protection, if trouble has started in the flock.

Navel ill is caused by bacteria which infect the navel of the new-born lamb before it has healed up. It is a common infection of young lambs, causing inflamed navel, hot and inflamed leg joints, dullness, lack of appetite, and death within 1–2 weeks.

Precautions and control

- Trouble to be expected if lambing in a place used for lambing before.
- Apply an antiseptic dressing or an aerosol to the navel as soon after the birth as possible.

Orf is a virus disease, sometimes found in late summer. It causes scabs and great irritation on mouth and nose of lambs, on ewes' udders and just above the hoof. It is difficult to cure. If trouble is expected, vaccinate the sheep.

Scrapie is a nervous disease, not usually found before 18 months of age. It is spread in breeding and is a rather mysterious complaint. It causes great irritation, rubbing, excitement and other nervous signs. Some breeds suffer more than others. It has been suggested that there is some link with BSE (see page 115).

Twin lamb disease (pregnancy toxaemia) is not a disease caused by any germ, but a disorder influenced by feeding and management. If in-lamb ewes are fed too well in early pregnancy and starved later, there will be trouble. The feeding should be adjusted to the real needs of the ewe and the lamb growing inside her.

Symptoms More commonly found in ewes carrying twins. Dullness, uncertainty, staggering, no appetite, sometimes blindness. Sometimes sweet smell on breath (acetone).

If the ewe can be kept going until she lambs, she will often recover.

Precautions and control

- Proper management and feeding of the pregnant ewe.
- Exercise.
- Extra feeding in last 6 weeks of pregnancy. Cereals, glucose and molasses can help.
- Dosing affected ewes with glycerine, glucose, Vitamin D and minerals sometimes helps.
- Put affected ewe in a pen, with supply of water, hay and cereals.

Swayback and pine are caused by deficiencies.

Swayback is caused by shortage of copper in lambs, and is shown by unsteadiness and swaying. *Pine* is caused by shortage of cobalt, and is shown by lambs doing poorly, with poor fleece, dull eye and weakness.

These troubles are found in certain districts; they are not general. The feeding of a mineral mixture to sheep is important, either in powder or block form. Sheep can also suffer from hypomagnesaemia (see page 118).

Parasites

- **External** Ticks, keds, and lice are sometimes troublesome, in addition to the mites that cause scab. All are controlled by dipping or by thorough spraying.

Fly is also less troublesome than it used to be. This can be expected in hot muggy weather, where there is plenty of cover (such as bracken). The green-bottle fly is attracted by moist dirty wool, particularly by scouring sheep, and lays its eggs in the wool. These hatch into grubs which quickly eat into the skin and flesh.

Symptoms Sheep appear unsettled and flustered, stamping feet, wagging tail, scratching and biting affected parts, standing about, looking miserable, dirty moist patches in wool, with maggots.

Precautions and control

- Keep sheep's hindquarters clean, by dagging if necessary.
- Keep worms under control, as worms cause scouring, etc.
- Treat affected sheep.
- Dip or spray. Modern chemicals give protection for 4 to 6 weeks.

- **Internal** Worms are normally found in sheep, and on average reduce their rate of growth by at least 15 per cent. This complaint is one of the chief reasons for the old saying 'A sheep's worst enemy is another sheep'.

Worms in stomach and intestines There are several different types, which live and breed inside the sheep, laying eggs which pass out in dung, develop into larvae which get on to grass and then infect the sheep again.

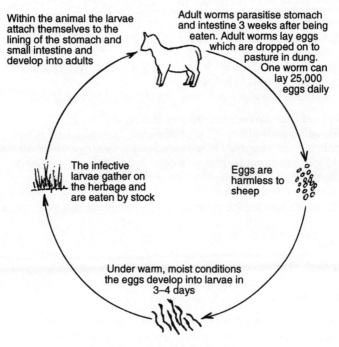

Within the animal the larvae attach themselves to the lining of the stomach and small intestine and develop into adults

Adult worms parasitise stomach and intestine 3 weeks after being eaten. Adult worms lay eggs which are dropped on to pasture in dung. One worm can lay 25,000 eggs daily

The infective larvae gather on the herbage and are eaten by stock

Eggs are harmless to sheep

Under warm, moist conditions the eggs develop into larvae in 3–4 days

Figure 35   Worms of sheep

Symptoms Scouring, dirty tails and hindquarters. Doing badly—the sheep do not grow well.

Precautions and control

- Move sheep onto fresh land, to avoid land becoming too much infected.
- Graze lambs especially on clean land (creep grazing can help).
- Treat ewes with worming medicines or injection about the time of lambing. First dose lambs at 6 weeks old and throughout the summer.

Liver fluke The fluke (flat worms) live in the sheep's liver where they feed, and produce eggs which pass out in the dung.

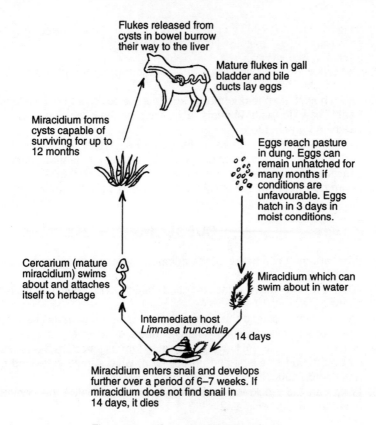

Figure 36 Life cycle of the liver fluke

The flukes pass into a small snail found in damp land, develop inside them and then pass to the liver where they develop.

Symptoms Poor condition, sheep becomes thin and miserable. Sometimes scouring, swollen belly, and swellings under chin.

Precautions and control

- Drain wet land.
- Control the snail by use of chemical (copper sulphate).
- Fence off wet places.
- Dose sheep with suitable medicines.

THINGS TO DO

1. Get to know the signs of health and vigour of all farm livestock and how to interpret these signs.
2. Learn how to take temperatures and feel the pulse of farm livestock.
3. Learn how to hold animals and restrain them so they can be examined and treated.
4. Learn how to inject animals, to take care of syringes, medicines, etc., and how to give medicines properly. Study the leaflet 'Code of Practice for the Safe Use of Veterinary Medicines on Farms'.
5. Learn how to nurse a sick animal and what to do until the vet comes.

QUESTIONS

1. What are the main needs of the animal to keep it healthy and in good condition?
2. What are the main signs of good health and condition in farm animals?
3. To what points should the farmer and stockman pay particular attention in order to keep livestock healthy and productive?
4. What are the main notifiable diseases of farm livestock at the present time? If found or suspected on the farm, what should you—and the farmer—do about it?
5. What can the stockman and the farmer do to control the common parasites of farm animals?

Chapter 7

Stockmanship

THE HEALTH, well-being and profitability of farm animals depend very largely on the care and observation of the man or woman who works with them—the stockman. This is a very important job, which is almost as much an art as a science, although it requires a thorough scientific understanding today.

LIVESTOCK JUDGING

Livestock Judging

Judging livestock can be a useful exercise—or it can be a rather pointless game. If it merely consists of running through a list of points which have little or no connection with commercial stock farming, then there is no reason for it. On the other hand, anything which makes you look at animals closely, and try to compare them, is worth while.

Avoid making a mere list of points, starting at the nose and finishing at the tail. First, look at the animals as a group, and then as individuals. Above all, remember the purpose the animal is being kept for—breeding or meat production—and judge it accordingly.

Listed below are the main points to look for; also given are scales of marking as used in some stockmen's competitions, which show the relative values of the different features.

Pigs Look for a long, lean body, light at the head and the fore-end; good ham development, and a reasonable depth to the body—giving a good side. Light bone is a good point. This means not too much bone for the size of the body—the more obvious bones (legs and shoulders) should not be too thick and heavy.

Remember which are the expensive parts of the carcase—the ham and the back—and look for good development there and as little as possible in the rest of the body.

Bacon Pig Judging—Maximum 100 points

Length	Ham and tail	Head	Shoulders	Belly and flank	Side
30	15	10	10	10	5
		Flesh and depth			
		20			

Dairy cattle First look at the whole shape of the body, which should be long and fairly deep at the rear. It should be a sort of triangle, deep at the back, narrow at the front, not too blocky. A good udder is important, reaching far up at the back and well forward under the belly. Strong straight legs are needed, as there is plenty of weight to carry.

Dairy Cow Judging—Maximum 100 points

Head and neck	Body	Legs and feet	Udder and teats
15	30	15	30

Character and temperament
10

Beef cattle The animal's body should be blocky, with short legs. The valuable meat is in the rear quarters and the back (roasting joints) and this is where you should look for good development. The meat should come well down the hind legs. You can judge a good beef animal entirely from behind—this is the part which matters most.

Beef Cattle Judging—Maximum 100 points

High proportion of meat to bone	High proportion of lean to fat	High proportion of meat on the roasting jonts
40	30	30

Sheep As a sheep is nearly always well covered with wool, feeling is often more important than seeing. By sight alone, you can see the nice long body of a good breeding ewe. Plenty of vigour and liveliness. Remember that she should be LONG, LOW and LUSTY.

By feel you can tell the good meat development, broad back, firm solid neck and plenty of meat development on the loin and around the tail. These are the points to look for in a good fat lamb or Down ram, which should have a solid blocky body.

Poultry There are a few simple signs of a good laying hen. Comb and wattles (fleshy skin below the throat) should be a bright red colour, not pale or shrunken. She should have a bright eye, and the beak should be straight, not crossed or broken.

Between the pelvic bones (two bones just in front of the vent) you should be able to feel about 3 fingers width; less than this and she is not laying much. There should also be plenty of 'capacity'—you should be able to measure a good hand's breadth between the end of the breastbone and the pelvic bones.

RECORDS

Some simple records must be kept of all livestock on the farm. Ministry of Agriculture returns have to be made each year, showing the numbers of animals of different types. Movement of stock on and off the farm must be recorded for disease control purposes.

More detailed records for business purposes, production standards and breeding are increasingly needed. There are ready-made systems including the use of computers.

Whatever else you keep, a day book or diary is essential. Never trust to memory—it will probably be wrong. Jot things down as they happen—the vet's visit, when the pigs were injected, deliveries of food, sales of stock—everything.

Production Records

Pigs There are recording schemes which help breeders and other pig farmers. Under these schemes you record the number of pigs born per sow, and the number alive at weaning time. Weights are recorded at various ages.

Pigs are commonly weighed during the finishing period; it is very important that bacon pigs should not be overweight towards slaughter as prices drop over certain weights. Quantities of food eaten by pigs are recorded and this can be related to their growth (see page 88). Grading for bacon and pork quality should be recorded—this information is available from the processors, also age at slaughter and carcase weights.

Cattle Calves can be weighed at birth and at various stages throughout their lives. Comparing these sets of figures shows the rate of liveweight gain and the progress made by each animal. Weighing beef animals also shows their progress and rate of gain, and relating these figures to food rations gives a conversion rate.

Dairy cows are commonly milk recorded. This can be done privately or under national milk records, which provide reliable information of use to breeders and buyers. Recording can be either daily or weekly, and the milk yield is recorded in kilograms. Milk is tested for quality (see page 19).

Sheep Simple sheep recording covers the numbers of lambs born to each ewe and the numbers weaned. More detailed recording includes weights at birth, at 8 weeks and 12 weeks, slaughter weights and ages. Weight of fleece per ewe can also be recorded.

Poultry In large units such as deep litter, only mass records can be kept—numbers of eggs per house per day—and this is related to the number of birds. Thus 500 eggs a day from a 1000-bird unit is a 50 per cent lay. In battery cages, individual records can be kept, usually on a card which is marked for each egg laid.

The food consumption of both layers and table birds is very important and accurate records must be kept. This information can be related to food used per dozen eggs or per kg of liveweight gain.

Breeding Records

A simple record of each female animal—sow or cow—shows date of service (mating), name of boar or bull, date of birth, number born, number reared and so on, during the whole life of the animal.

This information, along with the production records of the animal herself and her offspring, allows a proper breeding programme to be carried out. Details of any troubles or diseases suffered by the animal should also be recorded.

Some work of this type is done with sheep, but it is far more common to do it on a flock basis rather than for individual ewes.

Pedigree Records

Records of pedigree stock are kept by the breeder and are also recorded by the various breed societies. The societies are notified of the birth of a pedigree animal and give it an official number. Accuracy is important, because there must be true records of parentage.

HANDLING LIVESTOCK

The essentials in dealing with all animals are to be quiet, calm and deliberate. This makes handling them very much easier and pays in saving time and effort. There is nothing more infuriating than an animal which will not do what is wanted or go where it should, but this can be avoided in most cases by properly designed buildings, pens and other handling methods. Special crates, cradles and other handling devices are made for all types of stock. For pigs there are castration cradles; for cattle, crushes and holding pens and dehorning crates for calves; for sheep there are cradles and all sorts of catching and handling pens. If you have any number of stock to deal with, and particularly if you have to do any of this work single-handed, get the right equipment—and thus save a good deal of time and trouble.

A stockman should study his animals, know their likes and dislikes, have a good working knowledge of animal behaviour and psychology.

Along with this, there are some methods of handling stock which you should know:

Pigs will go where you want if you can make it difficult for them to see any other path than the one they should take. It is far quicker to use pig boards, or sheets of metal or hardboard to move pigs about, than to shout and wave your arms.

A grown pig can be restrained by putting a noose with a slipknot in its mouth and over the top jaw. Little pigs can be held easily with one hand or arm supporting them underneath and the other holding the head and ears. To open the mouth, press in with a thumb and finger on each side of the jaws.

Cattle Calves can be caught in a corner of the pen and a halter put over the head; this helps with handling for any purpose. Larger cattle may need to be haltered for leading, and this should be done quickly and firmly. For dosing or drenching, put one arm over the neck and with the other hand hold the nostrils firmly, raising the jaws and opening the mouth. Horned cattle can be held by the horns with one hand. To 'cast' (put down on the ground) an animal, putting a rope around the body a couple of times and pulling from the rear is very effective, but needs practice.

Sheep The one essential is to know how to 'turn' a sheep. This means putting one arm around and under the neck, and lifting with the other hand under the thigh. By this means, the sheep can be turned and set up on its backside, with its back towards you. In this position, it can be examined, treated in various ways, and held with very little effort.

A dog is necessary to round up the sheep in the field, and always makes handling easier. Sheep usually go upwards if they can, and towards other sheep, and these points are used in designing good sheep pens and buildings.

Young lambs are held with an arm around the body and the hand beneath the chest.

HOUSING LIVESTOCK

The housing of farm stock is important. Very often it is a case of making do with existing buildings on the farm, and although they cannot always be replaced if not satisfactory, they can be improved. There are a few general principles which apply to all buildings for housing stock.

Ventilation is needed to make sure of a supply of fresh air for the animals, and to take away the foul air. The more animals in a building, the more ventilation is needed. It is usual to give cattle (and sheep when housed) plenty of fresh air; they do not need to be kept very warm, although calves need a higher temperature than older cattle. Pigs and poultry need to be kept warm and it is common today to have controlled ventilation for them, with fans.

Used air usually goes out at the top of the roof, and fresh air comes in at some point in the walls. The inlets should be baffled, so that there are no direct draughts.

Insulation is extremely important for pigs and poultry and is often needed for calves. By using insulation material (fibre glass, plastic foam, etc.) heat is kept in the building. The roof is the most important place to insulate, followed by the walls.

Floors should always be insulated if animals lie direct on the floor—as pigs usually and calves often do. This can be done with hollow bricks, or various sorts of insulation material in the floor.

Damp-proofing the floors must always be done to protect against rising damp. This usually consists of a skin of bitumen or similar material under the top layer of the floor.

Drainage is needed in most livestock buildings, particularly if litter is not used on the floors. The best system is to slope the floor to gulleys which take the liquid away out of the building.

Disinfection of buildings should be done regularly between one batch of stock and another. This prevents the spread of diseases and parasites. For this reason, it is best to have floors, fittings and divisions which can be cleaned properly; in many ways concrete and metal are better than wood.

Cleaning out has to be done regularly. Where litter is used there may be a lot of material to shift. For this reason, it saves labour if tractors and trailers or sweeps can be used, instead of wheelbarrows and hand-work.

For some animals which are housed, there are ranges of temperature at which they do best. These are:

Piglets during first few days of life	21–26°C (70–80°F)
Piglets up to 4 weeks old	18–21°C (65–70°F)
Finishing pigs	12–18°C (55–65°F)
Calves during first 7 days of life	15–18°F (60–65°F)
Calves over 7 days old	10–12°C (50–55°F)

LIVESTOCK COSTS

Costs are all important in farming today. You must know which are the biggest costs in the production of an animal or any other product. Accurate records must be kept of these costs. In pig-keeping, where food accounts for 80 per cent of total costs, records of all food used are essential.

The control of some of these costs may be in your hands. Food is wasted by overfeeding, and usually this does no good at all.

STOCKMANSHIP

A good stockman must be observant; he must always be aware of what is going on with his stock. There are simple signs of health, which have been discussed in the chapter on each animal, but apart from these, farm animals show their health and general condition in many ways, and a good stockman will know at a glance if all is well—and if any trouble is to be expected.

Appearance of the animals is not the only indication of health and wellbeing. You can sense whether they are contented and doing well. Smell indicates some things—the first signs of some diseases and other troubles. Examples are the sweet smell of acetonaemia, the ripe cheesy smell of footrot, and the sour stink of calf scour. Different animals make their own sounds, and something can be told from this. Examples are the contented clucking of hens, the barking of good vigorous pigs and the steady chewing of the ruminant animals.

The stockman should be able to judge the growth of animals and the progress they are making—whether they are young stock growing or older animals fattening. He should be alert for any check in condition; this may be a sign of trouble brewing, or of faults in feeding or management.

Good stockmanship means keeping up to date. There are new developments in housing, feeding, management and disease control all the time—and new diseases and troubles to go with them. It is usually wise not to rush into new methods, but you should know what is happening, watch other people's experiments and be ready to take up anything new when it is proved in practice.

Read the farming magazines, discuss your problems and ideas with other farming people, and if possible join in the activities of a Stockman's Club.

THINGS TO DO

1. Learn to look for the signs of quality in livestock, and practise stock judging.
2. Get to know the ages of cattle and sheep by their teeth.
3. Learn how to clean and disinfect livestock buildings and equipment thoroughly.
4. From performance records and any other information, learn how to pick out the best animals for breeding.
5. Get to know the details of the welfare codes for farm livestock—and the farm safety regulations.

QUESTIONS

1. What is the body type and appearance of a good quality meat-producing pig?
2. What are the main points of a good quality dairy cow?
3. How do you tell the quality and readiness for sale of a good fat lamb?
4. What are the main production records which are kept for any type of farm livestock?
5. What are the main points that need to be considered in housing intensive livestock for modern conditions?

CONVERSION TABLES

BRITISH TO METRIC

LENGTH

1 inch (in) = 2.54 cm
          or 25.4 mm
1 foot (ft) = 0.30 m
1 yard (yd) = 0.91 m
1 mile   = 1.61 km

METRIC TO BRITISH

1 millimetre (mm) = 0.0394 in
1 centimetre (cm) = 0.394 in
1 metre (m)   = 1.09 yd
1 kilometre (km) = 0.621 miles

Conversion Factors

inches to cm × 2.54
     or mm × 25.4
feet to m   × 0.305
yards to m   × 0.914
miles to km × 1.61

centimetres to in   × 0.394
millimetres to in   × 0.0394
metres to ft   × 3.29
metres to yd   × 1.09
kilometres to miles × 0.621

AREA

1 sq. inch (in^2) = 6.45 cm^2
1 sq. foot (ft^2) = 0.093 m^2
1 sq. yard (yd^2) = 0.836 m^2
1 acre (ac)   = 4047 m^2
          or 0.405 ha

1 sq. centimetre (cm^2) = 0.16 in^2
1 sq. metre (m^2)   = 1.20 yd^2
1 sq. metre (m^2)   = 10.8 ft^2
1 hectare (ha)   = 2.47 ac

Conversion Factors

sq. feet to m^2 × 0.093
sq. yards to m^2 × 0.836
acres to ha   × 0.405

sq. metres to ft^2 × 10.8
sq. metres to yd^2 × 1.20
hectares to ac   × 2.47

VOLUME (LIQUID)

1 fluid ounce (1 fl oz)
  (0.05 pint) = 28.4 ml
1 pint   = 0.568 litres
1 gallon (gal) = 4.55 litres

100 millilitres (ml or cc) = 0.176 pints
1 litre   = 1.76 pints
1 kilolitre (1000 litres) = 220 gal

Conversion Factors

Pints to litres   × 0.568
gallons to litres × 4.55

litres to pints   × 1.76
litres to gallons × 0.220

WEIGHT

1 ounce (oz)   = 28.3 g
1 pound (lb)   = 454 g
            or 0.454 kg
1 hundredweight (cwt) = 50.8 kg
1 ton   = 1016 kg
            or 1.016 t

1 gram (g)   = 0.053 oz
100 grams   = 3.53 oz
1 kilogram (kg) = 2.20 lb
1 tonne (t)   = 2204 lb
            or 0.984 ton

Conversion Factors

ounces to g   × 28.3
pounds to g   × 454
pounds to kg   × 0.454
hundredweights to kg × 50.8
hundredweights to t × 0.0508
tons to kg   × 1016.0
tons to tonnes   × 1.016

grams to oz   × 0.0353
grams to lb   × 0.00220
kilograms to lb   × 2.20
kilograms to cwt × 0.020
tonnes to tons   × 0.984

Index

Farming Press Books

The Principles of Dairy Farming   Ken Slater

An introduction, setting the husbandry and management techniques of dairy farming in its industry context.

Calculations for Agriculture & Horticulture
Graham Boatfield & Ian Hamilton

Provides the calculation methods for crops, livestock, machinery and horticulture.

Farm Machinery   Brian Bell

Gives a sound introduction to a wide range of tractors and farm equipment. Now revised, enlarged and incorporating over 150 photographs.

Farm Workshop   Brian Bell

Describes the requirements of the farm workshop and illustrates the uses of the necessary tools and equipment.

For more information or for a free illustrated catalogue of all our publications please contact:

Farming Press Books & Videos, Wharfedale Road Ipswich IP1 4LG, United Kingdom Telephone (0473) 241122   Fax (0473) 240501

and Videos

Farm Crops   **Graham Boatfield**

A basic introduction to farm crops and crop husbandry.

Calving the Cow and Care of the Calf
Cattle Ailments
Pig Ailments   **Eddie Straiton**
Sheep Ailments

These handy volumes are profusely illustrated and packed
with straightforward, practical advice.

An Introduction to Keeping Sheep
                                  Jane Upton & Dennis Soden

The skills and techniques of caring for sheep for newcomers.

Poultry at Home (VHS colour video)
                                          Victoria Roberts

A beginner's guide to poultry health and management.

Farming Press Books & Videos is part of the Morgan-Grampian
Farming Press Group which publishes a range of farming magazines:

*Arable Farming, Dairy Farmer, Farming News, Pig Farming, What's New
in Farming*. For a specimen copy of any of these please contact the
address on the facing page.